AIChE Spring Meeting 2005

DOE / ITP Funding of Novel Technologies 2005

Held April 10-14, 2005
Atlanta, Georgia, USA

Printed from CD-ROM with permission by:

Curran Associates, Inc.
57 Morehouse Lane
Red Hook, NY 12571
www.proceedings.com

ISBN: 978-1-60423-466-4

T4 - DOE/ITP Funding of Novel Technologies

Session 21 - Novel Reaction Technologies R&D Funded by U.S. DOE/ITP

Chair: Joseph V. Porcelli
Vice Chair: Jo Rogers

1 **Membranes for Corrosive Reactions**
Jamie A. Hestekin, Michael P. Henry, Seth W. Snyder and Yupo J. Lin

2 **Low Temperature Electrochemical Ammonia Synthesis**
Adrian Denvir, Kyle Uselton and Oliver Murphy

3 **Nanostructured Polymeric Heterogeneous Catalyst for Industrial Applications**
Brian J. Elliott and Douglas L. Gin

4 **Direct Conversion of Alkane to Valuable Chemicals - Gas-phase Selective Oxidation of n-Butane to Methyl Ethyl Ketone**
D. Strongin, F. Dautzenberg, J. Hao, K. Pillai, L. Li, M. Lin, M. Zubic, N. Burm, P. Jain, V. Chilukuri and X. Wang

5 **Production and Separation of Fermentation-Derived Acetic Acid**
David J. Law, Edward J. St. Martin, Kelly A. Skinner-Nemec, Mark Donnelly, Michael P. Henry, Mike Muskett, Milind Patel, Rathin Datta, Robert J. Niedzielski and Seth W. Snyder

6 **Methane to Methanol Conversion – Compatibility study of Using Ionic Liquids as Novel Reaction Media**
William A. Goddard III

7 **Biphasic Hydroformylation of Higher Olefins**
Steven Dietz and Trudy Sholten

Session 22 - Novel Separations and Enabling Technologies R&D Funded by the U.S. DOE/ITP

Chair: Joseph V. Porcelli
Vice Chair: Jo Rogers

12 **Low Emission Diesel Engines**
Don Stookey and Stuart Nemser

41 **Thin Film Ceria Membranes for Oxygen Separation**
Edward M. Sabolsky, Matthew M. Seabaugh and Scott L. Swartz

54 **Novel Polymer Nanocoating Processing to Produce Ceramic Nanocomposites**
Alan Weimer, John Ferguson, Joseph Spencer, Karen J. Buechler and Steven George

55 **Novel Air-VOC Chemically Inert Reverse Separation Membranes**
J. Bowser, S. Nemser and R. Grantham

99 **Hollow Fiber as Structured Packing for Olefin/Paraffin Separation**
Craig Colling, Dali Yang, David Devlin, E. L. Cussler, Martin E. Carrera and Robert Barbero

Session 23 - Miscellaneous Enabling Technologies R&D Funded by the U.S. DOE/ITP

Chair: Joseph V. Porcelli
Vice Chair: Jo Rogers

21a Membranes for Corrosive Reactions

Jamie A Hestekin, Michael P Henry, Seth W Snyder, Yupo J Lin

Hydrogen peroxide is widely used in pulp and paper applications, environmental treatment, and other industries. Virtually all hydrogen peroxide production is now based on a process featuring catalytic hydrogenation followed by auto-oxidation of suitable organic carrier molecules. This process has several drawbacks, particularly in the extraction phase. One general disadvantage of this technology is that hydrogen peroxide must be produced at large centralized plants where it is concentrated to 70% purity by distillation, and then transported to the users' plant sites where it is substantially diluted before use.

Advanced membranes have the potential to enable more efficient, economic, and safe manufacture of hydrogen peroxide. Advanced membrane technology would allow filtration-based separation to replace the difficult liquid-liquid extraction based separation step of the hydrogen peroxide process. This would make it possible for hydrogen peroxide to be produced on-site in mini-plants at working concentrations and used in situ or at the same plant location without distillation and transportation.

Industrial partners on this project are UOP LLC (Des Plaines, IL) and Unitel Technologies (Mt. Prospect, IL)

21b Low Temperature Electrochemical Ammonia Synthesis

Adrian Denvir, Kyle Uselton, Oliver Murphy

Approximately 3% of the world's energy is consumed by industrial synthesis of ammonia, and in many developing countries the capability for ammonia synthesis is the first sign of budding industrialization. The most important method of manufacturing ammonia is the Haber-Bosch process, which accounts for over 90% of the world's ammonia production. However, even under optimum reaction conditions, yield of ammonia is low and the overall system suffers from extreme inefficiencies. Lynntech has developed an electrochemical ammonia generation system based on the oxidation and reduction of nitrogen containing species in a molten salt. Preliminary results show that ammonia can be generated at lower temperatures and pressures than the Haber-Bosch process.

21c Nanostructured Polymeric Heterogeneous Catalyst for Industrial Applications

Brian J Elliott, Douglas L Gin

Polymeric acid catalysts are increasingly used in the chemical industry. However, current polymeric heterogeneous catalysts are amorphous, and do not have well defined, regular pores with controlled diameters and tight pore size dimensions (as do, for example, high temperature zeolite acid catalysts). It is now known that if acid groups can be ordered on the inside of 10 to 15 nanometer diameter pores (as the active sites) results in the catalysis of reactions with higher activity, fewer side reactions, and at lower temperatures. TDA Research and the University of Colorado are developing a process for making polymeric acid catalysts by a self-assembly route. The resulting nanostructured catalysts performance is superior to conventional polymeric catalysts because they can carry out size or shape selective reactions. These catalysts are being evaluated for industrial chemical processes including esterification, phenol alkylation and other acid catalyzed reactions.

21d Direct Conversion of Alkane to Valuable Chemicals - Gas-phase Selective Oxidation of n-Butane to Methyl Ethyl Ketone

D Strongin, F Dautzenberg, J Hao, K Pillai, L Li, M Lin, M Zubic, N Burm, P Jain, V Chilukuri, X Wang

Methyl ethyl ketone (MEK) is a highly demanded commodity chemical with an annual production in 2002 of nearly 700 million pounds in the US and 2.2 billion pound worldwide. Currently, 87 % of the MEK production is via a three-step sec-butyl alcohol (SBA) process, which starts with 1-butene, consumes large amounts of energy, uses corrosive sulfuric acid, and generates massive amounts of corrosive and toxic wastes. This SBIR research project, awarded to EverNu Technology LLC by the DOE, aims at exploring and developing a cost and energy-saving, as well as environmentally friendly catalytic process to produce MEK in one step from n-butane, a cheap and abundantly available alkane with air as the oxidant. The objective of the Phase I SBIR research was proving the technical feasibility of producing MEK directly from n-butane in a one-step gas-phase catalytic oxidation using air as the oxidant. The mission of the current Phase II research is to achieve significant n-butane conversion and MEK selectivity concurrently, leading to an attractive one-pass MEK yield for further development and commercialization, and to conduct preliminary study on the economic feasibility of the proposed new technology, as compared to the current conventional SBA process.

21e Production and Separation of Fermentation-Derived Acetic Acid

David J Law, Edward J St. Martin, Kelly A Skinner-Nemec, Mark Donnelly, Michael P Henry, Mike Muskett, Milind Patel, Rathin Datta, Robert J Niedzielski, Seth W Snyder

Acetic acid is a large volume chemical product and the United States is the world's largest producer of virgin acetic acid. Currently, 80 percent of the U.S. production of virgin acetic acid is produced by methanol carbonylation. However, this synthetic process requires relatively high temperatures and pressures, a toxic/corrosive environment, exotic materials of construction, extensive safety-related equipment. The result is high capital cost and low scaling factor, which makes only very large plants economically feasible (~1000 million pounds per year). At these plant sizes, significant distribution networks are required, and transportation costs become a large factor in the contract selling prices. Therefore, there are strong market, economic, and energy benefits to developing processes for production of acetic acid in scalable, regional-sized plants, such as fermentation. The advantages of producing acetic acid by fermentation include its appropriateness for small-scale production, lower cost feedstocks, low energy membrane-based purification, lower temperature and pressure requirements, and decreased transportation costs with regional plants.

One of the existing barriers with anaerobic fermentation is that anaerobic acetogens are most productive at near neutral pH. The team is developing an integrated fermentation-derived production and purification process that addresses the barriers of acidification, cation recycling, and economic recovery of the acid.

21f Methane to Methanol Conversion – Compatibility study of Using Ionic Liquids as Novel Reaction Media

William A Goddard III

Interactions of methane and methanol with various ionic liquids in the presence of concentrated sulfuric acid and/or Pt-based catalysts are investigated both experimentally and theoretically. As preliminary efforts of developing the low-temperature and highselectivity CH4-to-CH3OH homogeneously catalyzed conversion system, solubility and stability between the oxidant (H2SO4), the catalyst ((bpym)PtCl2) and the media (ammonium, phosphonium, imidazolium, and pyridinium based ionic liquids) have been systematically studied. In addition to those commercially available ionic liquids, a series of new media with modified cations are designed and synthesized, showing improved stability in strong acid. Fundamentals of the interaction mechanisms are further understood using molecular simulation techniques at the atomic level.

Biphasic Hydroformylation of Higher Olefins

Steven Dietz and Trudy Scholten

TDA Research, Inc.
12345 W. 52nd Ave.
Wheat Ridge, CO 80033

Introduction

Currently, worldwide production of aldehydes exceeds 7 million tons/year (Billig and Bryant 1995). Higher aldehydes are important intermediates in the synthesis of industrial solvents, biodegradable detergents, surfactants, lubricants, and other plasticizers. The process, called hydroformylation or more familiarly, the Oxo process, refers to the addition of hydrogen and the formyl group, CHO, across a double bond.

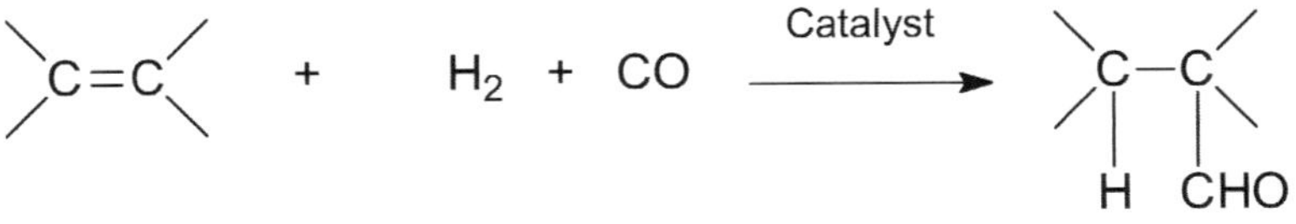

For lower aldehydes, homogeneous rhodium-phosphine catalysts are used because they have by far the best activity and selectivity. In this case the aldehyde product is recovered by distillation before the catalyst is recycled. The process used for conventional hydroformylation of short-chain alkenes cannot be applied to higher (>C_6) olefins due to the difficulty of recovering the high-boiling product aldehyde without simultaneously destroying the catalyst. The hydroformylation of higher olefins (>C_6) is done with less efficient cobalt catalysts at high temperatures and pressures.

One approach that has been successfully used to separate the catalyst from the product aldehyde is to use a biphasic system in which the rhodium catalyst is soluble in water and the product soluble in an organic phase. This approach is used by Hoechst/Rhône-Poulenc to produce more than 600,000 t/year of butyraldehyde (a lower aldehyde) (Baker and Tumas 1999). Unfortunately, this process cannot be used to produce higher aldehydes because the water solubility of the higher olefins that are the feedstock is very low, which dramatically reduces the reaction rate.

To eliminate the need to recover the product by distillation, thermomorphic solvent mixtures are proposed. These mixtures are immiscible at room temperature but become miscible at reaction temperature, enhancing conversion rates by promoting contact between the rhodium catalyst and the olefin. After the reaction is complete, the reaction mixture is cooled and the phases completely separate. The product can be simply recovered by decantation and the catalyst can be recycled. Thermomorphic mixtures can result in dramatic energy and cost savings because the fraction of feedstocks converted to the desired product could be increased from 60 to 85%.

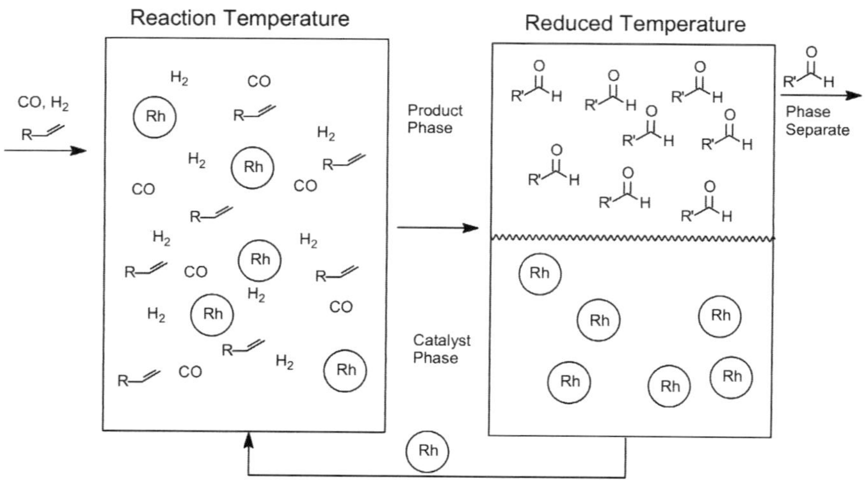

Figure 1. Representation of hydroformylation in a thermomorphic system.

A thermomorphic system is characterized by solvent pairs that reversibly change from being biphasic to monophasic as a function of temperature. Many solvent pairs exhibit varying miscibility as a function of temperature. For example, methanol/cyclohexane and *n*-butanol/water are immiscible at ambient temperature, but have consolute temperatures (temperatures at which they become miscible) of 125°C and 49°C, respectively (Halpern and Gozashti 1988).

The use of thermomorphic systems has recently been studied as a way of achieving catalyst separation in homogeneous catalysis. For example, a biphasic hydroformylation catalyst system was developed to take advantage of the unusual solvent characteristics of perfluorocarbons combined with typical organic solvents. Fluorous mixtures such as perfluoromethylcyclohexane (PFMCH)/toluene) are immiscible at ambient temperature but become a single-phase solution upon heating (Hildebrand and Cochran 1949; Scott 1948; Dorset 1990). This characteristic is very useful for homogeneous catalysis because the reaction can be performed at an elevated temperature where the mixture becomes one-phase, overcoming any rate limitations due to mass transfer. Upon reaction completion, the solution is phase separated by cooling and the products are simply decanted. Unfortunately, the expense of the solvents and the ligands make the fluorous biphasic system impractical for industrial application. Other thermomorphic systems studied rely on exotic phosphine ligands for their use. Again, these are not used industrially because of their high cost (Bergbreiter et al 1998, Bianchini et al 1995).

Our approach is to use the inexpensive ligands that are already used industrially as well as conventional solvents. The goal of this project is to develop a thermomorphic approach to rhodium-catalyzed hydroformylation of higher olefins (>C_6) that enhances conversion rates and ease of product recovery while minimizing catalyst degradation and loss.

Thermomorphic solvent mixtures have been tested for hydroformylation of 1-octene, 1-dodecene and tetradecene to determine the ease of product recovery and catalyst recycling. Using a small scale Parr pressure reactor, we demonstrated the efficacy of a biphasic, thermomorphic, system that had the following advantages:

-Suitable for conversion of higher olefins to higher aldehydes.
-High conversion rates and selectivity
-Recovery of the product aldehyde by simple room temperature decantation rather than high temperature distillation.
-Little to no catalyst leaching and high catalyst recyclability.
-More than 30% saving in feedstock and energy costs.

Results

In order to do long term studies, we have built a bench-scale continuous hydroformylation apparatus. Figure 2 is a picture of the apparatus built at TDA and Figure 3 depicts the P&ID for the apparatus. The apparatus is fully automated using OptoControl software and is designed to operate unattended. It has an Autoclave Engineers reactor with a 300 ml capacity that features a bolted closure with a confined gasket for high temperature sealing. It is designed to withstand a maximum allowable pressure of 3800 psig at 538°C. The contents of the reactor are mixed with an overhead mixer designed specifically for use with the pressure vessel. Two thermocouples measure the internal and external temperature of the reactor wall and a pressure gauge monitors the internal reactor pressure. A rupture disk designed to withstand up to 1000 psig is mounted on the top of the reactor. All process lines to and from the reactor are unheated 1/8" 316 SS tubing with a wall thickness of 0.3".

Figure 2. Bench-scale continuous hydroformylation apparatus.

The reactor is a continuous stirred reactor (CSTR) with a catalyst residence time of approximately 30 minutes. Two separate 2.1L reservoirs contain the catalyst and product phases and the contents are fed into the reactor through a standard liquid mass flow controller. Under reaction conditions, (100°C and 500 psig $CO:H_2$) the phases become miscible and hydroformylation occurs. The contents of the reactor can be sampled from a pressure fed sample tube. The pressurized liquid reactor products exit the reactor through a pressure control valve, which reduces the pressure to atmospheric, and the liquid contents are delivered to a continuous decanter where the phases separate. The catalyst phase then settles to the bottom where it is drained for recycle and reuse, while the product phase is collected into a 4.2L reservoir.

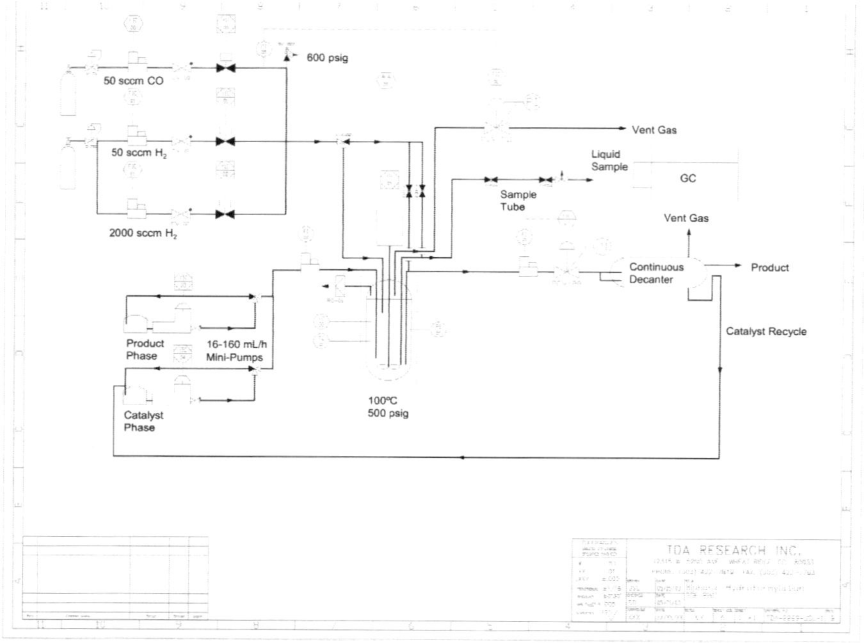

Figure 3. P&ID of bench-scale continuous reactor.

In Figure 4 is shown an example of the hydroformylation of 1-octene in the continuous bench-scale reaction using thermomorphic catalysts. The reaction was carried out with 1:1 molar mixture of H_2/CO with a combine pressure of 500 psi at 100°C for over 350 hours. Samples were taken from the reactor and the product analyzed by GC. The 1-octene conversions averaged 50% at the current flow rate. We believe the scatter in the data is due the drift in the pump flow rate, which alters the residence time and not the catalyst itself. In all cases the linear to branch aldehyde selectivity was very high in the range of 4-5:1 linear to branched aldehyde. We have found that the rhodium loss to the product phase is very low, below 1 μ/g as measured by atomic absorption. At this time we believe the rhodium loss due to leaching of catalyst into the product phase is minimal and should not be a problem.

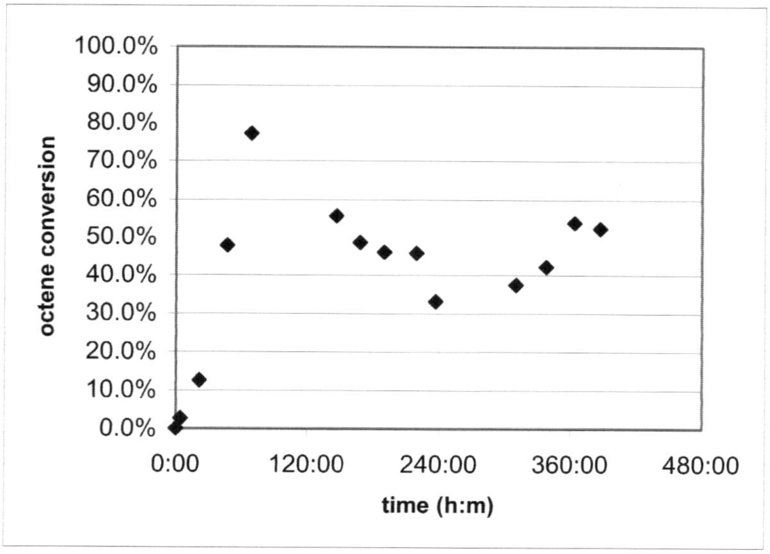

Figure 4. Percent octene conversion vs. time

Now that we have shown that the system works using 1-octene as the higher olefin, we will begin testing the higher olefins 1-dodecene and 1-tetradecene. From the results of the tests in this continuous reactor system, we hope to have data strong enough to convince potential commercial partners of the advantage of our process compared to conventional cobalt-based hydroformylation of higher olefins.

Conclusions

We have developed a thermomorphic catalyst system for the hydroformylation of higher alkenes. We have built a bench-scale continuous reactor and are using it to determine the long-term performance of the thermomorphic catalyst system. Initial results using 1-octene show that the activity and selectivity of the catalyst are high and the rhodium losses are very low. The next step is to demonstrate the process with the higher olefins dodecene and tetradecene.

Acknowledgment

This work is being supported by the Department of Energy under SBIR Phase II grant DE-FG03-01ER83307/A002.

References

Baker, R.T., and W. Tumas (1999). "Toward Greener Chemistry," *Science*, **284**, 1477-1479.

Bergbreiter, D.E., Y.-S. Liu and P.L. Osburn (1998). "Thermomorphic Rhodium(I) and Palladium(0) Catalysts," *J. Am. Chem.. Soc.*, **120**, 4250-4251.

Bianchini, C. P. Frediani, and V. Sernau (1995). "Zwitterionic Metal Complexes of the New Triphosphine $NaO_3S(C_6H_4)CH_2C(CH_2PPh_2)_3$ in Liquid Biphasic Catalysis, *Organometallics*, **14**, 5458-5459.

Billig E. and D.R. Bryant (1995) "Oxo Process," in Encyclopedia of Chemical Technology, 17, 902-919.

Dorset D.L. (1990). "Binary Phase Behavior of Perfluoroalkanes," *Macromolecules* **23**, 894-901.

Halpern, A.M., and S. Gozashti (1988). Mutual Solubilitites of Liquids in a Binary Two-Phase System," J. Chem. Educ., 65,371-372.

Hildebrand, J.H., and D.R.F. Cochran (1949). "Liquid-Liquid solubility of Perfluoromethylcyclohexane with Benzene, Carbon Tetrachloride, Chlorobenzene, Chloroform and Toluene," *J. Am. Chem. Soc.* **71**, 22-25.

Horváth, I.T., and J. Rábai (1994). "Facile Catalyst Separation Without Water: Fluorous Biphase Hydroformylation of Olefins," Science 266, 72-75.

Scott R.L. (1948). "The Solubility of Fluorocarbons," *J. Am. Chem. Soc.* **70**, 4090-4093.

Project Title: Low Emission Diesel Engines

Key Personnel: Stuart Nemser, Ph.D. PI
Don Stookey, Ph.D.

Participating Organizations:

- Argonne National Lab
- Caterpillar
- DuPont

NOx Emissions Reduction

- NOx, oxides of nitrogen, formed during combustion
- Particularly high temperature combustion
- Reducing combustion temperature reduces NOx
- Dilution of engine cylinder charge with inert gases lowers temperature and thereby reduces NOx
- Possible diluents
  - nitrogen
  - carbon dioxide
  - exhaust gases
  - water vapor or steam

The CMS approach …

- Nitrogen Enriched Air (NEA)
- from cooled turbocharged air
- with an air separation membrane
- integrated into the diesel engine and
- preferably on-board diesel powered trucks and mobile equipment

Diesel Engine with NEA Membrane

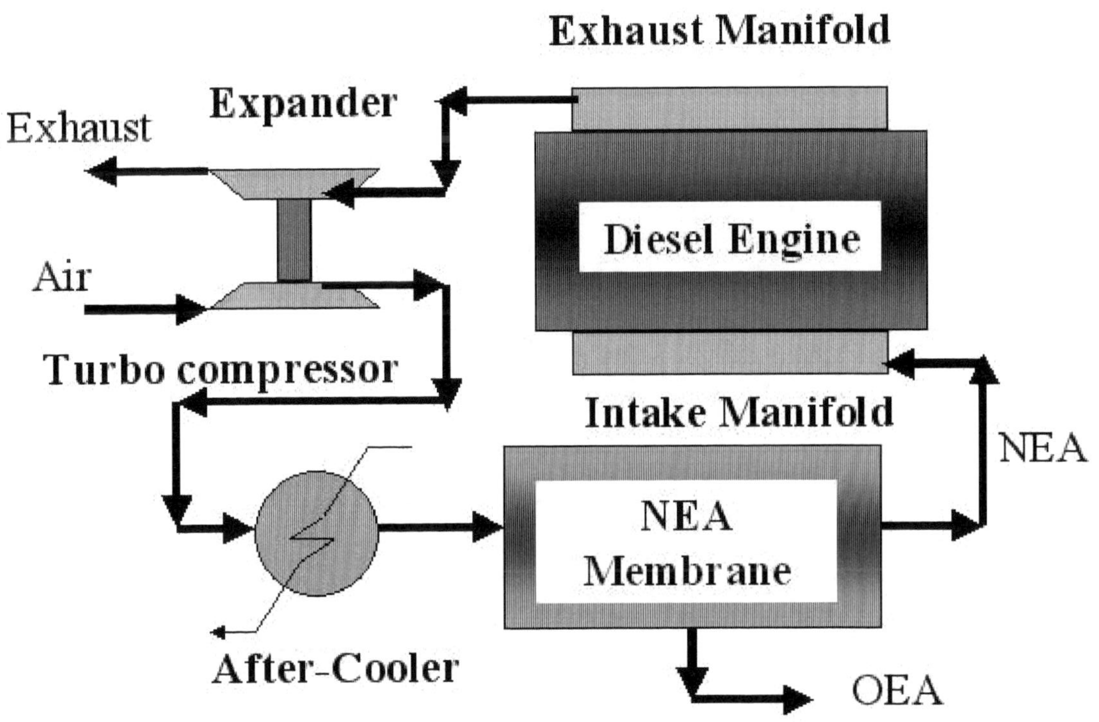

AIChE April 2005

PROGRAM ACTIVITY – NEA-NOx

- CMS DEVELOPED NEA FOR NOx RED.
  - CLEAN, COOL EGR
  - NO HEAT LOAD ON ENGINE
  - NO WEAR ON ENGINE
  - EASY RETROFIT
  - WORKS BEST AT HIGH LOAD
  - SULFUR IN FUEL NOT AN ISSUE
  - 50% REDUCTION IN NOx

Key Accomplishments

1) Optimized Membrane (25%)
2) Identified/Developing Large Opportunities (50%)
3) Patented Sweep for Improved Design (25%)
4) Major Overall Improvements Additive
   - A) 72% reduction in mem. Size/cost (1,2,3)
   - B) 63% reduction in energy/need (2,3)

Key Accomplishments (Cont'd)

5) Active participation of Caterpillar (+ data)

6) Detailed durability results (e.g. 1,000,000 mile tests)

7) Additional field test planned

8) Caterpillar showcased at DOE DEER

NEA Membrane in Engine Compartment

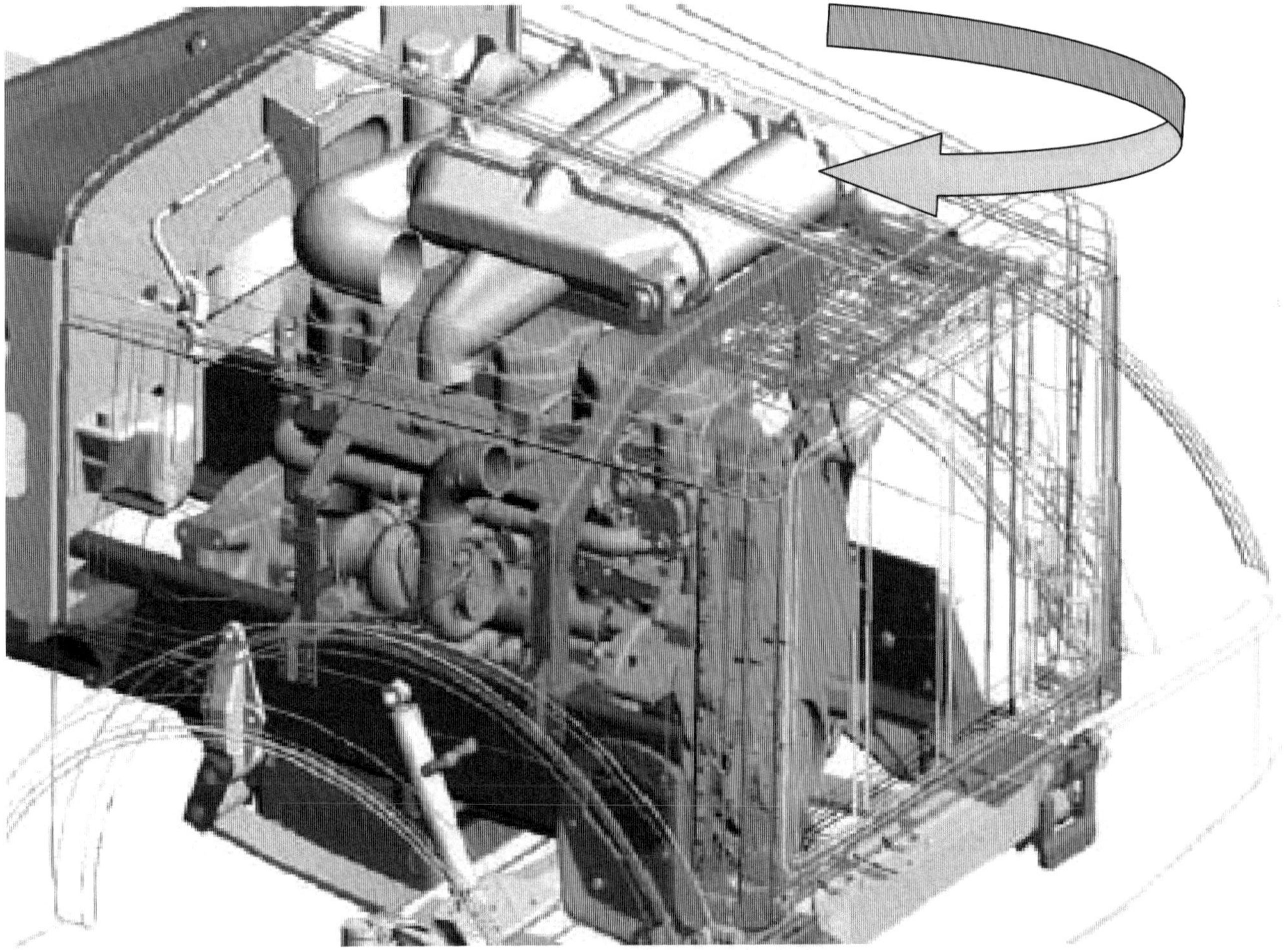

AIChE April 2005

Membrane Housing above Engine

AIChE April 2005

Reality - Membranes Under the Hood!

AIChE April 2005

IMS Hollow Fiber Membrane Cartridge, 6”dia. x 17”

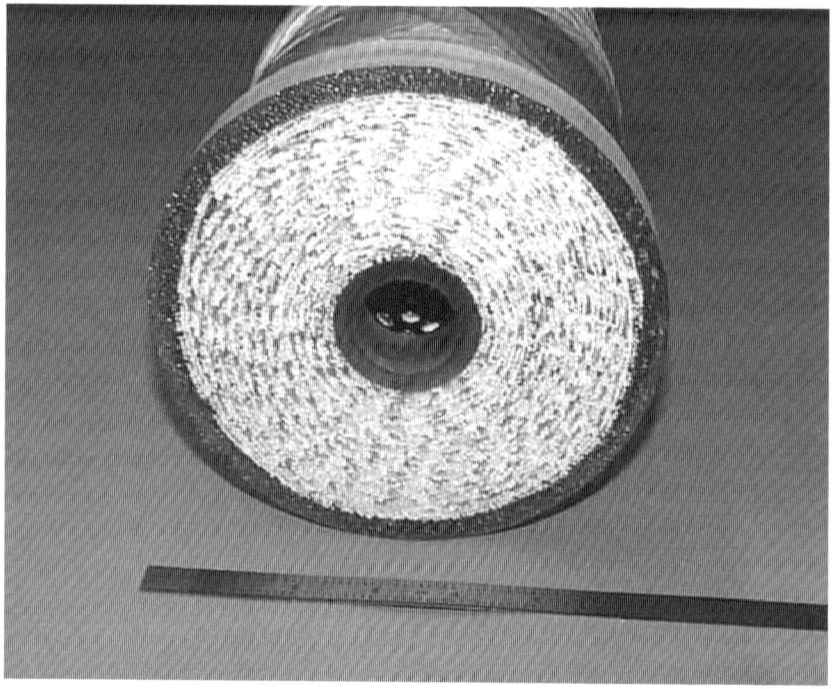

Durability Tests

- Laboratory Testing
  - Accelerated Durability Tests
    - 300,000 pressurization cycles
  - High Temperature Aging Effects
    - Membrane properties with cycling
  - Road Dust Test
  - Emissions Tests on Laboratory Engine
- Field Tests

Life at High T – Keeps ON …

Long-Term NEA Cartridge Performance

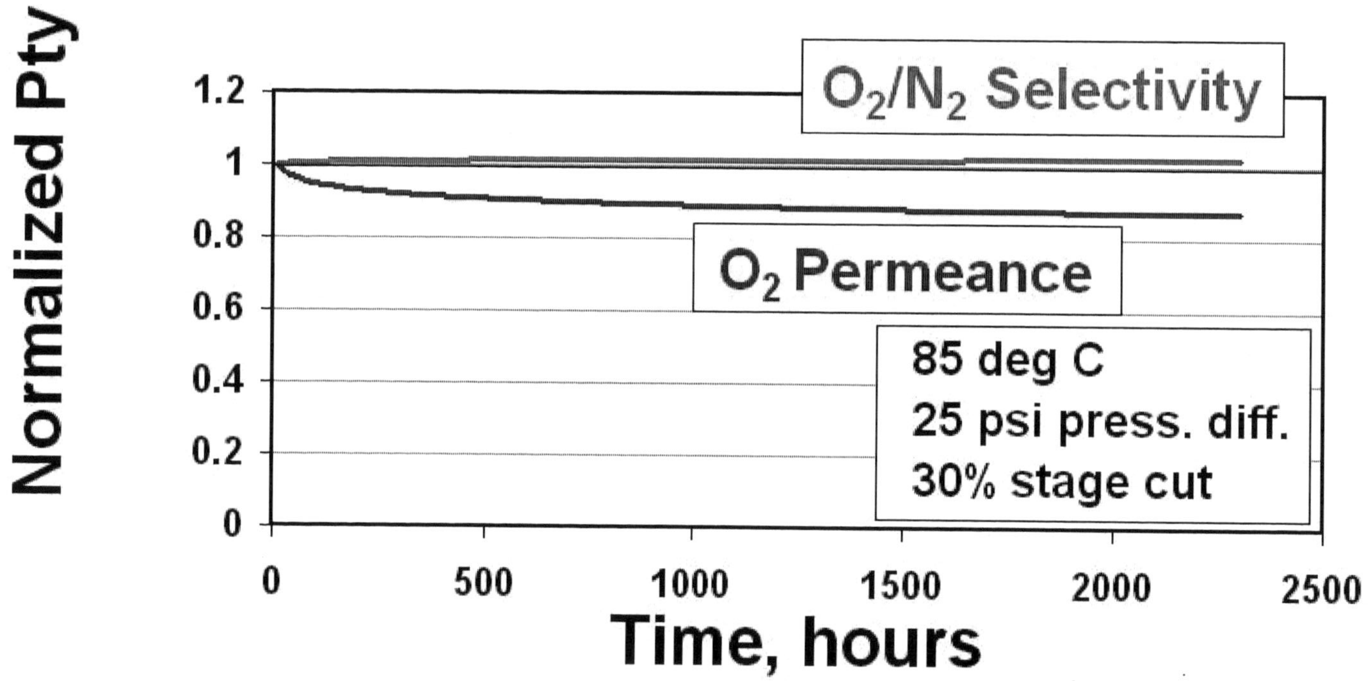

AIChE April 2005

SwRI Accelerated Road Dust Test

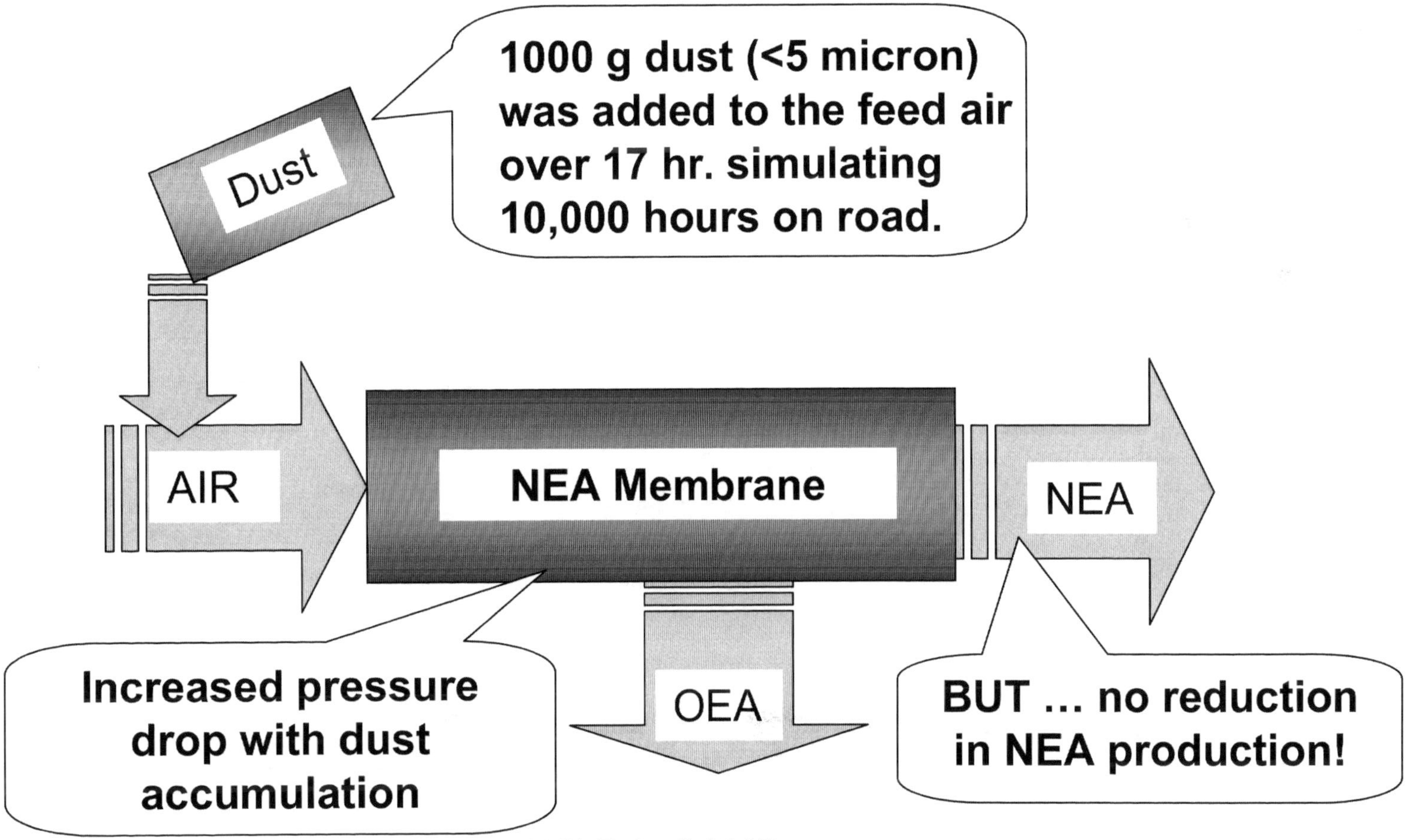

AIChE April 2005

Lab Engine Emissions Tests

Steady State 13 Mode Cycle Results on NEA Modified CAT C-12 Engine

(g/hphr)	2000 6g NOx Production C-12 Engine	C-12 Engine With NEA	2004 EPA Standard
NOx	5.5	2.41	
NOx + HC	5.6	2.50	2.50
Particulate	0.04	0.097	0.100

Field Durability Test Program

- Five Class 8 Highway Trucks
- Outfitted with NEA Membranes
- Approaching **1 million miles** on the road
- No evidence of Membrane Fouling
  - All Five Over 150,000 miles
  - One over 220,000 miles

Five Highway Trucks

- Over 150 K mi. each, One over 200 K mi.
- One evidenced air leak, removed for inspection

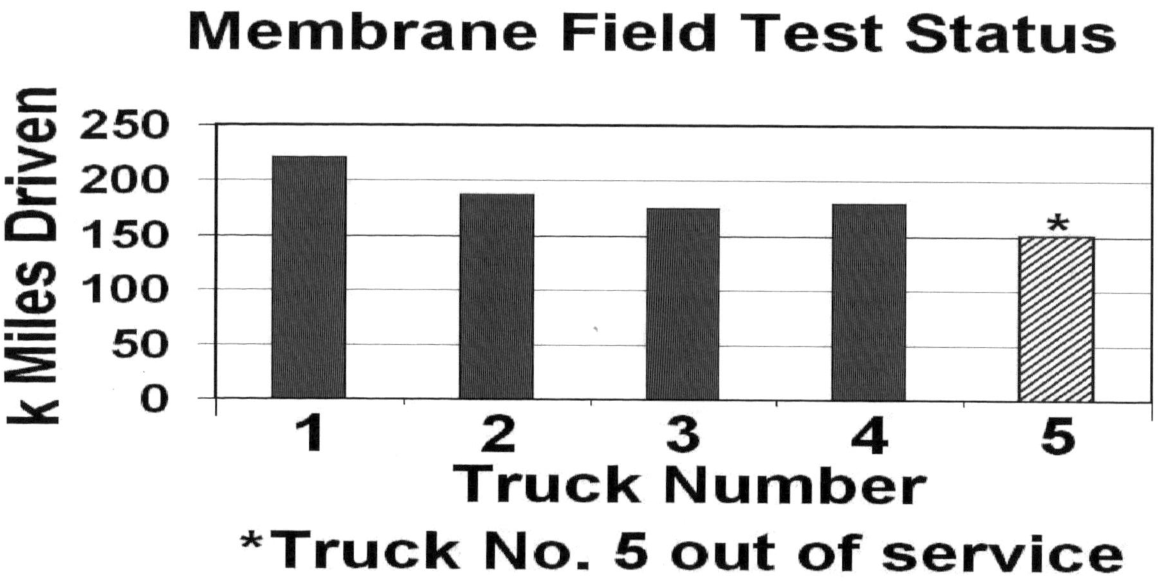

AIChE April 2005

CMS' Partners and Supporters

- DuPont
  - Licensor to CMS
  - Perfluorinated Polymer Supply

Argonne National Lab
  - Engine Research with OEA and NEA
- Caterpillar, Inc.
  - Application Development
  - Demonstration
  - Field Testing

AIChE April 2005

Key Issues

- Performance vs. Market Need
  - Cost
  - Durability
  - Energy
- Market
  - On-road
  - Locomotive / Marine

Membranes need Driving Force

- Locomotive / Trucks 30 psi
- Passenger Car 5-10 psi
- Hybrid Diesel 12 psi

Demonstration of Sweep Enhanced System Performance

	Mem.Sel. (O_2/N_2)	Retentate ($\%N_2$) No Sweep	Retentate ($\%N_2$) Sweep
Simulation	2.5	81.6	82.7
Actual Membrane	2.63	81.4	83.0

No Sweep

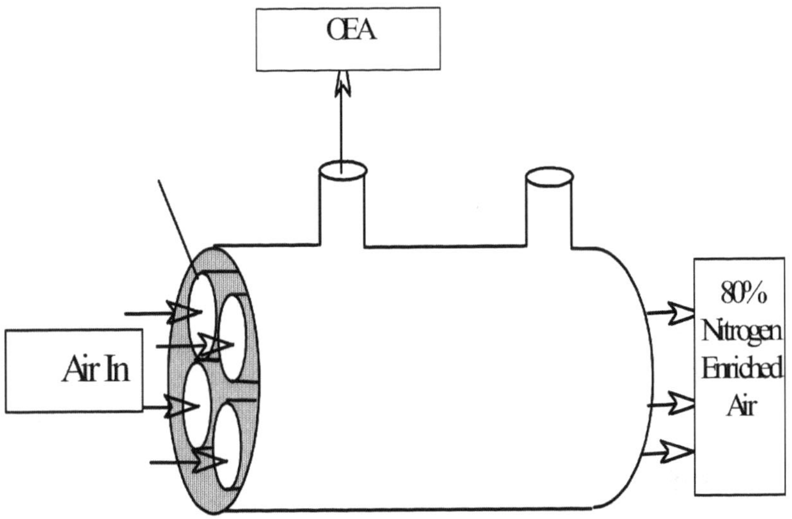

Air Sweep

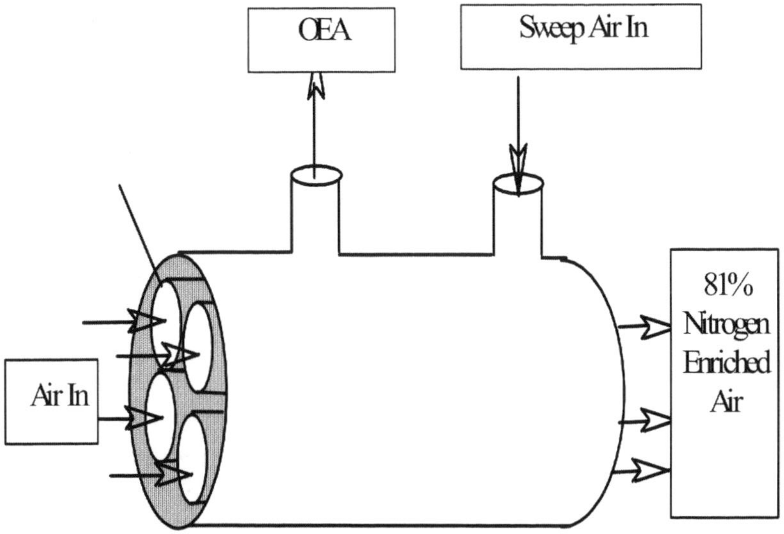

Exhaust Emissions Standards for Locomotives*

	Gaseous and Particulate Emissions (g/bhp-hr)			
	HC	CO	NOx	PM
Tier 0 line-haul duty-cycle	1	5	9.5	0.6
Tier 1 line-haul duty-cycle	0.55	2.2	7.4	0.45
Tier 2 line-haul duty-cycle	0.3	1.5	5.5	0.2
Tier 0 switch duty-cycle	2.1	8	14	0.72
Tier 1 switch duty-cycle	1.2	2.5	11	0.54
Tier 2 switch duty-cycle	0.6	2.4	8.1	0.24

Recent Emissions Pressures*

- Spotlight on Maritime
- IMO Annex 6 regulations
- EPA regulations 1,2, & 3 engines
- CARB reqmt ferries use hwy diesel fuel in CA
- Ozone Nonattainment Areas – Coastal harbors
- Public Pressure
- Norway – Marine sources of Acid Rain

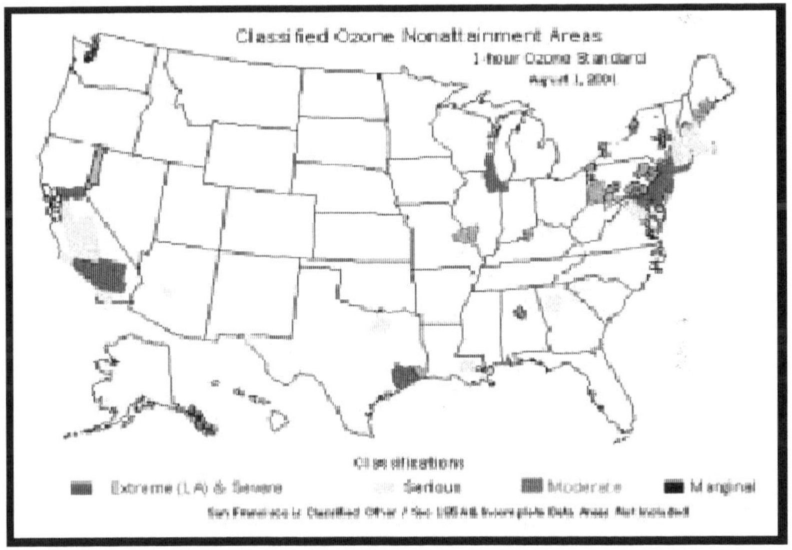

US Ozone

Nonattainment Areas

*Daniel J Gore, Maritime Administration, presentation at DEER 2003

AIChE April 2005

Maritime NOx Emitters

- Harbor District
  - Fishing Fleet
  - Tugboats and Cargo Ships
  - Ferries, Cruise, and Pleasure Craft
  - US Navy – permitted access?
- Dockside
  - Cargo handling equipment
  - Shore power connect vs. on-board generation in port
  - Cruise Ship, Navy, etc. in dock
- Coastal Boundary Waters
  - Fishing Fleet
  - Navy and Coast Guard
  - Cargo and Cruise Ships

Marine and Shipboard NEA

- Diesel Powered
  - Medium and Low Speed Engines
  - High horsepower ratings
- Unregulated Fuels
  - Refined diesel – local fleet
  - Bunker fuel – ocean ships
  - CNG – Norwegian Ferry
  - Widely varying sulfur content
- NEA Application = Retrofit of existing marine engines

Energy Impact

- NEA alone, ~ 4% BSFC* penalty at 40% NOx reduction
- However, the added degree of freedom allows changes e.g. timing, for BSFC savings with modest NOx reduction

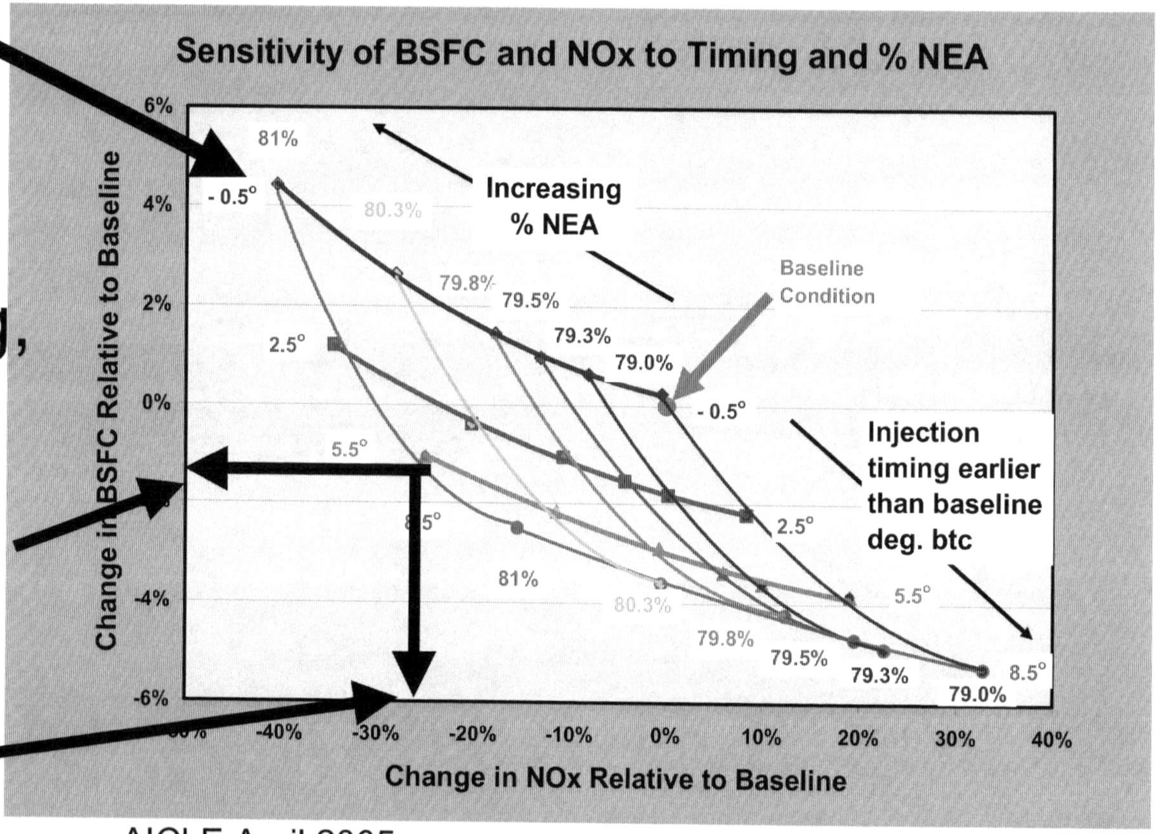

AIChE April 2005

* BSFC = Brake Specific Fuel Consumption, gm fuel/bhp-hr

Energy Impact

- Thus, the energy impact is engine protocol and application specific and tightly coupled with the NOx emissions reduction required.
- OEMs and operators are considering use of NEA in combination with other Emissions Reduction Technologies.

Energy Savings
Annual Fuel Savings

	Fuel Used (B gal)	Share (%)	BSFC Savings (%)	Fuel Saved (MM gal)	Energy Saved (Quad)
Diesel Engines*	46	10%	1%	46	0.01
Passenger Car Hybrid Diesel	150	0.1%	40%	60	0.01
					0.02

* Farm, Power, Rail, Vessel Bunkering, On-Highway, Off-Highway, Military

AIChE April 2005

Conclusions

- Working with Caterpillar demonstrated membrane system capabilities & limitations
- Excellent manufacturing (MEDAL) partnerships in place
- Initial commercialization focus shifted to locomotive / marine
  - High load
  - Steady operation
  - Significant opportunity
- Other longer term opportunities exist
  - Hybrid diesel passenger car
  - On-road / Off-road diesel

Acknowledgement

The authors gratefully acknowledge the support of the **US Department of Energy** through SBIR Contract DE-FG02-01ER83167

Doped-CeO_2 Thin-Film Ceramic Membranes for Small-Scale Oxygen Generation Systems

Edward M. Sabolsky, Ph.D.
Matthew M. Seabaugh, Ph.D.
Libeth Maldonado
Willie Collins
Scott L. Swartz, Ph.D.
NexTech Materials, Ltd.
404 Enterprise Drive
Lewis Center, OH 43035

Abstract:
NexTech Materials, Ltd. is developing a ceramic oxygen generation (COG) systems to meet the critical market need for generators which can provide pure (99.9+%) oxygen at pressures and rates satisfactory for small oxygen consumers who rely on cylinders or dewars. COG systems operate by oxygen ions being transported through a ceramic membrane, which is set between a pair of gas permeable (porous) electrodes. NexTech's SBIR program, through Department of Energy, focuses on the fabrication of a ceramic COG elements which are composed of a porous ceramic electrode tube coated with a dense Gd-doped CeO_2 (GDC) thin film electrolyte and a porous ceramic outer electrode.

Over the recent program period, porous ceramic electrode tubes were fabricated by extrusion and served as the mechanical support and the current collector for the cell. A modified lanthanum strontium ferrite (LSF) ceramic was chosen to provide the support, based on thermal expansion and electrical conductivity data. The composite LSF/GDC interlayer and GDC electrolyte layers (both about 10-20 μm thick) were deposited by aerosol-spray methods, and the coated tubes were then co-sintered to achieve the desired morphology. The spray deposition and co-sintering of the membrane active layers aim to simplify cell fabrication and reducing manufacturing costs. Also, research identified a range of ceramic electrode and composite electrode materials based on the lanthanum strontium ferrite (LSF) composition that demonstrate enhanced low-temperature interfacial resistance which show great promise for further enhancing the performance of the COG element when incorporated into the cells.

Introduction:
High purity oxygen is a high valued chemical product, part of a $15 billion oxygen market. While the raw material for oxygen production is free, the separation of air is expensive. For users requiring less than 1000 m^3/hr, the most economic means of delivery is to distill O_2 by cryogenic processes at a central location and distribute it in high-pressure cylinders or cryogenic dewars. Transportation of oxygen requires many precautions and costs are significant. Transportation is also energy inefficient — a 125 cubic foot oxygen cylinder, used in welding and medical applications, is 15% by weight oxygen and 85% metal. Both producers and consumers would benefit from lower costs if a high-purity distributed oxygen generators could be installed and transportation energy consumption eliminated. If on-site technologies are developed, high-purity industries (i.e. semiconductor, pharmaceutical, and health-care) could benefit from the lower cost, convenience, and stability of on-site oxygen production.

The industrial gas market has seen a steady decline in the demand for packaged gases, and increases in the delivery of bulk liquid dewars and on-site generation. This has been driven by the high additional costs of packaging and transporting gas cylinders and development of lower purity on-site solutions such as PSA and VPSA. If equivalent high-purity oxygen generation systems were available, with low operating costs, consumer demand would drive the new system to supplant packaged and bulk delivery for this market. At the smaller-scale, users of high-purity oxygen generators such as home health patients, hospitals, and metal fabricators would benefit from significant convenience, portability, and safety advantages over cylinders and dewars. As the technology is introduced and production costs reduced, one also can expect the increased use of oxygen in leading edge applications. There are several industrial combustion and chemical production processes where the advantages (efficiency, pollution abatement, etc.) of using pure oxygen instead of air are well established, but where costs of existing oxygen delivery have been prohibitive.

Ceramic oxygen generation systems for separating oxygen from air employ a ceramic solid electrolyte, and operate as shown in Figure 1. Oxygen ions are transported through the ceramic membrane, which is sandwiched by a pair of gas permeable (porous) electrodes. When a voltage is applied to the electrodes, an electric field is generated which causes electrical current to flow through the ceramic membrane. This electrical current flow is due to the transport of oxygen ions through the membrane: oxygen molecules from air are converted to oxygen ions at one electrode (cathode), oxygen ions then being transported through the membrane via oxygen vacancies in the crystal structure of the ceramic electrolyte to the opposite electrode (anode), where oxygen ions are converted back to molecular oxygen and collected. Successful development of a ceramic oxygen system requires careful consideration of the electrolyte and electrode materials, as well as the electrolyte element configuration.

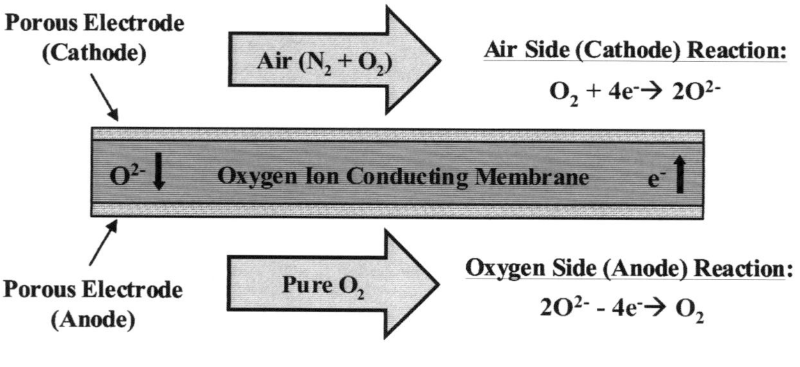

Figure 1. Operating Principle of Electrically Driven Ceramic Oxygen Generation

Thin Film Configuration and Processing:

There are several design approaches that have been considered for ceramic oxygen generation systems. Early work focused on building stacks of planar electrolyte elements or bundles of ceramic electrolyte tubes, with appropriate electrode coatings. Ceramic oxygen systems based on these relatively simple designs have been demonstrated in the laboratory, but have not been commercialized for several reasons. The relatively large thicknesses of the electrolyte membrane (~200 to 500 microns) and the use of low-conductivity electrolyte materials, like yttrium-stabilized zirconia (YSZ), make these systems electrically inefficient and require that the systems operate at high temperatures (~800°C and higher). Technical obstacles that have been encountered with such systems include thermal management, sealing and manifolding, electrical interconnection, long-term reliability, and volumetric efficiency. A number of creative design approaches have been applied to solve many of the above issues, including the conception and demonstration of novel monolithic electrochemical cell designs. Another emerging COG design approach is an electrode-supported thin-film which dramatically reduces the electrolyte thickness (<50 μm), and reduces power requirements for generating oxygen. The electrode-supported cells can be operated at lower temperature and at lower power than electrolyte supported systems, while achieving the same oxygen output. Current fabrication routes for these cells are expensive. With lower cost materials and fabrication technologies, such as those being pursued in this SBIR program, significant improvements over state-of-the-art systems can be achieved.

This SBIR program is focusing on fabricating a ceramic COG element which is composed of a porous ceramic electrode tube coated with a dense, thin Gd-doped CeO_2 (GDC) electrolyte and a porous ceramic outer electrode (Figure 2). The porous ceramic tubular support is fabricated by extrusion and serves as the mechanical support and the current collector for the cell. A modified LSF ceramic has been chosen to provide the support, based on thermal expansion and electrical conductivity data. The composite LSF/GDC interlayer and GDC electrolyte layers (both about 10-20 μm thick) will be deposited by aerosol-spray methods, and the coated tubes then will be co-sintered to achieve the desired morphology. This tri-layer approach has been successfully demonstrated at NexTech in DOE-funded research aimed at solid oxide fuel cell applications. The final LSF/GDC composite electrode layer also will be deposited by aerosol-spray followed by annealing.

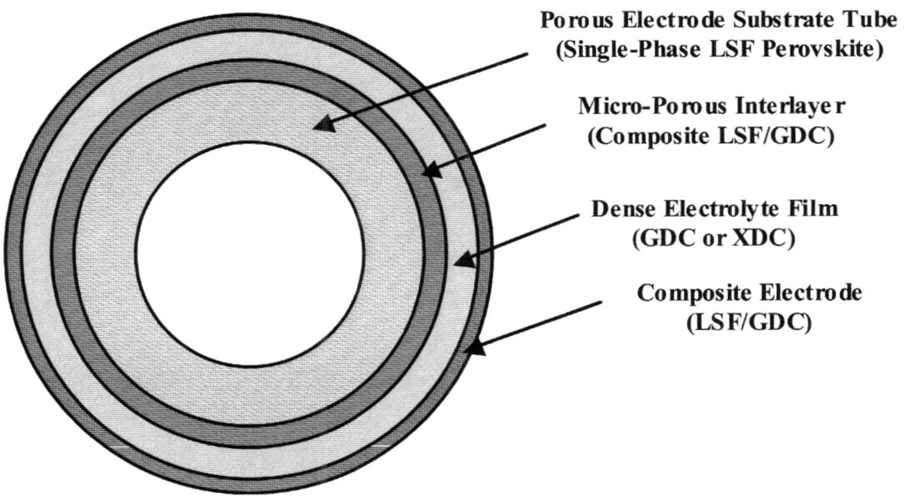

Figure 2. Cross-section of tubular electrode-supported, thin-film electrolyte element.

Materials Development:

CeO_2 electrolyte compositions were explored in this program by improving on various formulations previously studied by other researchers. NexTech has allocated various resources to develop inexpensive synthesis routes for manufacture doped-CeO_2 compositions that display high strength and conductivity. A co-precipitation method has been implemented to prepare doped-CeO_2 compounds (XDCs) utilizing inexpensive and commonly available raw materials. Single-phase crystalline compounds (XDCs) have been derived via calcinations of co-precipitated metal hydroxides. After various milling routes, the resulting powders show submicron particle size and an average particle diameter of less than 0.50 microns. Depending on the precipitation and milling process, the surface area of the XDC powders can be distinctly controlled and manipulated. NexTech has developed processes for producing powders with surface areas ranging between 2-50 m^2/g from anywhere between 1-20 kg. batch sizes. Some XDC compositions evaluated have a magnitude higher ionic conductivity than YSZ at 600°C and sintering temperatures that are up to 200°C lower than YSZ. The powder synthesis and processing routes allowed for low temperature processing of high-density and fine-grained ceria-based ceramic electrolytes, which resulted in ionic conductivity values that were higher than those previously reported in the literature.

In this program, the GDC composition was used as the baseline electrolyte composition in order to develop the aerosol spray deposition and co-sintering processes. The GDC powder was processed in the same manner described above for the various XDC powder compositions. In order to characterize the electrical and mechanical properties of this powder, the GDC powder was isostatically pressed into 2.5-cm pellets and sintered at 1300°C and 1400°C. Four-point DC electrical conductivity measurements were completed on test bars cut from these pellets, and the samples demonstrated conductivities between 0.006-0.04 S/cm at temperatures between 500-800°C, respectively (Figure 3). The samples sintered to 1400°C demonstrated equi-biaxial flexure strength of 75.5 MPa. This same GDC powder was utilized in the spray-deposition trials for the fabrication of the COG elements.

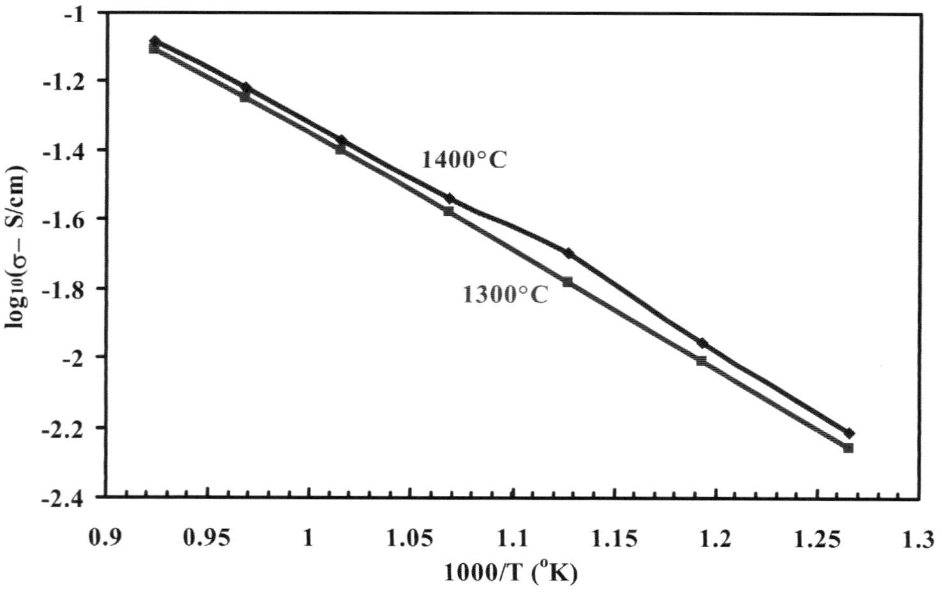

Figure 3. Arrhenius plot of the conductivity of dense GDC ceramic sintered at 1300°C and 1400°C. The GDC powder was synthesized by NexTech.

A range of ceramic electrodes and composite electrode compositions have also been investigated. The sintering performance, electrical conductivity, interfacial resistance of B-site doping of the lanthanum-strontium-ferrite (LSF) compositions was the focus of recent studies, in order to improve kinetics of the oxygen reduction and release at the electrodes. For improved kinetic performance, the electrode composition must show a high degree of electrical and oxygen ion conductivity near to the interface with the GDC electrolyte. At distances away from the interface, a high electrical conductivity is required in order to move electrons to and from the interface, which greatly reduces the chance of polarization at the interface. Therefore, the conductivity of the electrode interlayer and the electrode contribute greatly to the overall resistance of the cell, and thus, the cells efficiency.

$(La_{0.6}Sr_{0.4})FeO_3$ (LSF), doped-LSF (10 mol% Mg, Zn, Ni, and Cu), and doped-LSF/GDC composites were investigated in this work. Electrode compositions were prepared by a mixed-oxide or mixed-carbonate method, where the lanthanum, strontium, iron, and B-site dopant precursors were attrition-milled and then calcined at temperatures 1000-1300°C for 4 h to achieve phase purity. All compositions adopted the perovskite crystal structure, similar to LSF. The composite compositions were formed by later attrition-milling the GDC composition (35 vol%) with the doped-LSF powder previously synthesized. The phase pure powders were formed into an ink and were symmetrically screen-printed onto GDC disks. The thin films were sintered at temperatures between 1200-1400°C to adhere the layers onto the GDC substrate, yet leave the substrates with ~30-45% porosity. The electrode interfacial resistance measurements were measured by collecting the impedance spectroscopy using a Solartron 1260 on the symmetrical samples at temperatures ranging between 450-700°C. For comparison, the complex impedance curves were zeroed to the electrolyte/lead resistance (R_l) by simple interpolation. Figure 4 shows the impedance data collected at 600°C for the LSF and doped-LSF compositions. The overall area specific resistance values displayed in Figure 4 are quantitatively high due possibly to unknown impedance additions by the leads and fixtures, but the trends can be used to qualitatively show the performance difference between the various compositions. The figure shows that the Zn, Cu, and Mg samples show the lowest relative interfacial resistance, at least 2.5 times less than the standard LSF composition, with the Zn and Mg dopants (which exhibit the 2+ valence state) having particularly low values overall. This indicates that the (Cu, Zn, Mg)-doped LSF compositions showed enhanced mixed-conductive character at the GDC electrolyte interface, perhaps by the increase of oxygen reduction kinetics. LSF composition doped with Ni displayed an increase in interfacial resistance compared to the un-doped composition. These materials could provide very useful intermediate layers for the assembled cells. Interfacial resistance between the electrolyte and the electrode create some of the largest resistances in an electrolytic cell of this design. Reducing this effect will enhance oxygen flux and/or reduce power consumption. Additional benefits of the evaluated materials are that the thermal expansion values for the materials are within the range of acceptability for the ceria based electrolyte system.

Figure 5 displays the trend of the interfacial area specific resistances versus temperature for all of the doped-LSF compositions. The figure shows that the relative difference between the resistances for each of the doped-LSF compositions remain consistent over the entire temperature range. The (Cu, Zn, Mg)-doped LSF compositions still show the lowest interfacial resistance with the GDC substrate. The graphs also shows that the magnitude of the interfacial resistances for these compositions further decrease by at least fifty percent by increasing the temperature to ~700°C. Thus, this work shows that the overall resistance contribution of the electrodes to the COG element can be drastically reduced by substituting some of these B-site doped LSF for the standard LSF composition.

Figure 6 shows the impedance spectroscopy results for some of the same doped-LSF compositions shown in Figure 4, but some of the doped-LSF/GDC composite compositions were included for comparison. Figure 6 shows that the doped-LSF/GDC composites showed even further decrease in the interfacial resistance, again possibly due to the increased mixed-conductive character resulting from the homogenous LSF/GDC mixture. The LSZF/GDC and LSCuF/GDC showed at least a 0.8 ohm-cm^2 decrease in area specific resistance across the interface. Further research must be completed on the doped-LSF/GDC composites (with varying GDC ratios) to identify the optimum composition that provides the lowest cathode interfacial polarization.

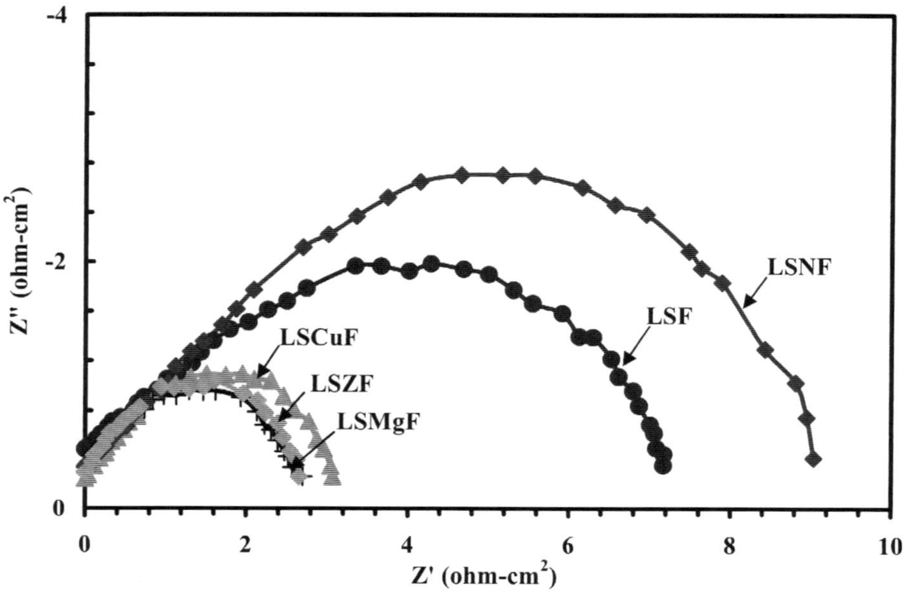

Figure 4. Electrode interfacial resistance (polarization) measured by impedance spectroscopy at 600° with symmetrically printed LSF and doped-LSF (10 mol% Mg, Zn, Ni, and Cu) on GDC disks.

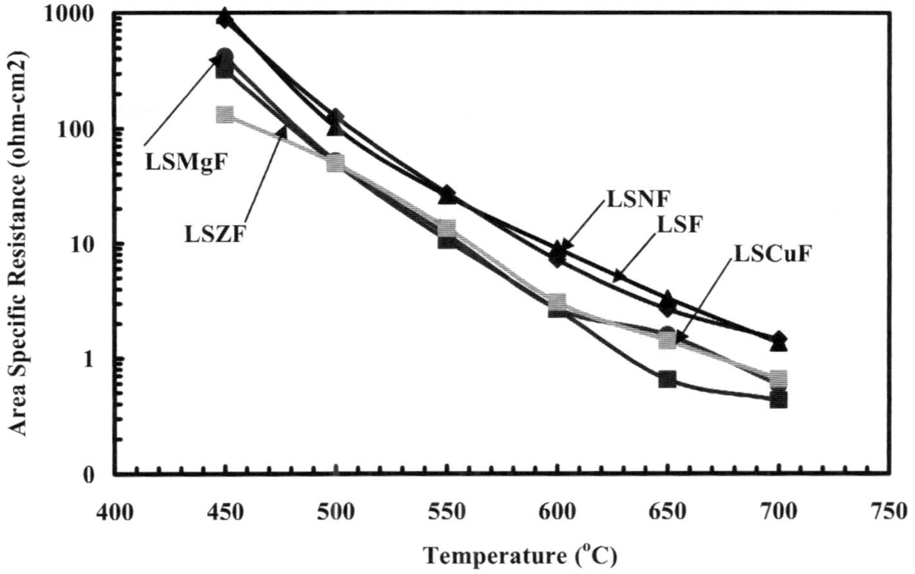

Figure 5. Area specific resistance versus temperature of LSF and doped-LSF (10 mol% Mg, Zn, Ni, and Cu) electrode compositions on GDC disks measured at 600°C.

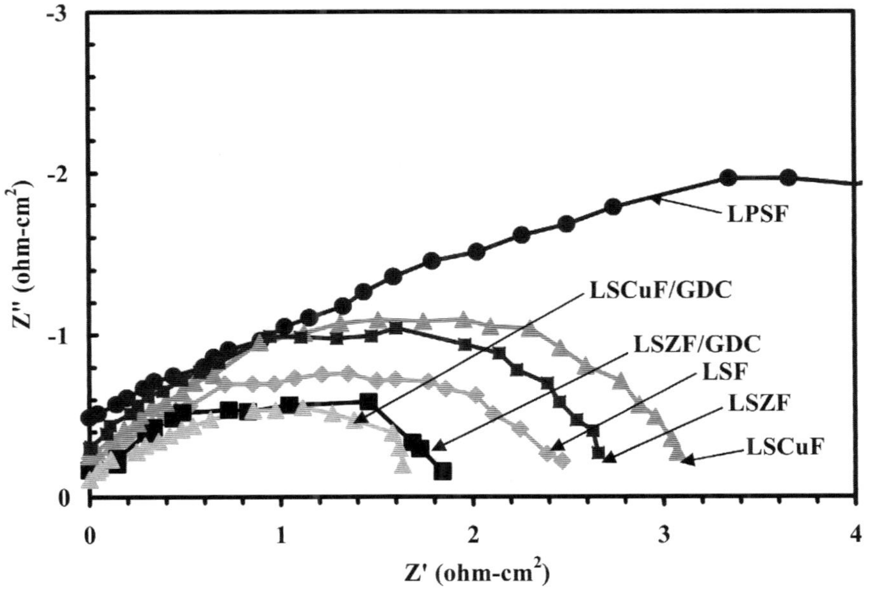

Figure 6. Electrode interfacial resistance (polarization) measured by impedance spectroscopy at 600° with symmetrically printed LSF, doped-LSF (10 mol% Mg, Zn, Ni, and Cu) and doped-LSF/GDC composites on GDC disks.

Cell Manufacturing:

The primary focus of this project is the fabrication of tubular, ceramic COG elements. As discussed above, these COG elements consist of a porous mixed-conductive ceramic tube coated with a dense, thin film, ceramic electrolyte layer. The outer surface of the element is painted or sprayed with a mixed-conductive ceramic layer, similar in composition to the tubular support. The LSF composition was used for the initial development of the electrode tubular support and outer electrode. As described above, the LSF composition, even without further B-site doping, displays high electrical conductivity (at lower temperatures) and relatively low interfacial resistance with the GDC electrolyte composition. The GDC composition was used as the basic electrolyte composition due to its enhanced oxygen conductivity even at lower usage temperatures, like the 550-700°C temperature range required for the COG application. These compositions were used during the initial COG fabrication trials in order to minimize any additional complications relating to chemical inhomegeneity or incompatibility. The use of these basic compositions provide a straightforward path for the development of the processing methodology and after full characterization of COG elements produced with these baseline compositions are completed, optimization will be carried out using the developmental electrode compositions discussed above.

The porous LSF supports were fabricated by an aqueous extrusion process. The LSF powders were first synthesized through a mixed-oxide route. The LSF powder was dry-mixed with organic fugitive material (pore former) and the organic binder. The required water was slowly added to the powder mixture in a sigma-blade mixer, after which, the dough was extruded through a Loomis 14-ton piston extruder. The water content of each batch was modified to produce a shear viscosity of ~40,000-90,000 Pa·s at a shear rate of 1 s^{-1}. LSF tubes were extruded to a length of ~60-80-cm and permitted to dry in a humidity controlled atmosphere. The long LSF tubes were then segmented in the green-state to a length of ~12.5-cm in length, and the final green tubes showed diameter of 1.4-cm in diameter and a wall thickness of 2.0-mm. The samples were then capped with green material and burned-out. Before coating the tubes with the electrolyte and outer-electrode layers, the tubes were pre-sintered to temperatures between 1000-1400°C in order to match the co-sintering shrinkage of the active layers and the tube.

The dough formulation, fugitive content, and LSF particle size distribution were developed in order to produce a tube which displayed a porosity level between 25-40%, an oxygen permeability of 0.015-0.022 cm^2/sec, and an electrical conductivity of ~100-140 S/cm at 700°C (porosity level ranging between 25-35%). Figure 7 shows the electrical conductivity as a function of temperature of both a dense (96% dense) and porous (75% dense) LSF tube measured by a four-point conductivity method. The LSF powder utilized in this work showed a baseline conductivity ranging between 250-300 S/cm between 450-700°C. The graph shows that the included porosity increased the resistive path across the material, causing the conductivity to drop to 140-160 S/cm between 450-700°C. The measurements for both the dense and porous LSF samples similarly show a gradual decrease in conductivity with an increase in temperature above 700°C.

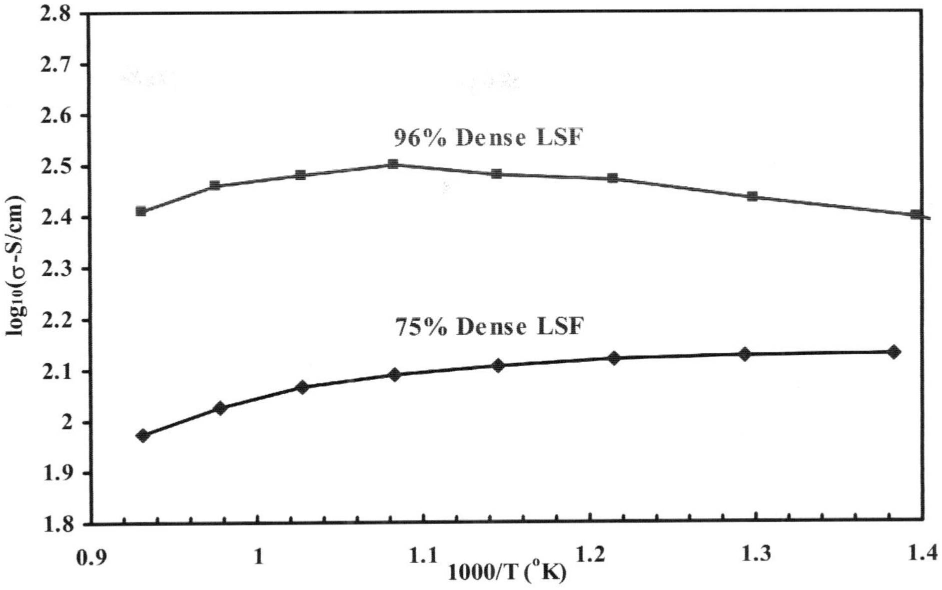

Figure 7. Electrical conductivity as a function of temperature for both a dense and porous LSF tubular sample.

Aerosol spray deposition is a low cost process which offers great flexibility for manipulating the chemical and microstructural makeup of the deposited layer. With aerosol spray deposition, a range of densities and pore distributions can be deposited through the control of particle size. In this project, spray suspensions were tailored for depositing each active layer onto the LSF electrode tube. The goal of the studies was to achieve the best deposition characteristics and the greatest control over sintering behavior and coating morphology. Process control was achieved by tailoring the particle size and surface area of the component powders. Figure 8 displays the automated aerosol spray deposition setup. The figure shows an ultrasonic spray-head attached to a motor which systematically moves parallel to the LSF tube. The spray suspension is supplied to the head by a varistaltic pump. The tube is attached along its center axis and is rotated at a rate ranging between 80-150 rpm. The coating thickness is primarily controlled by the suspension loading, the suspension flow rate, and the number of deposition passes.

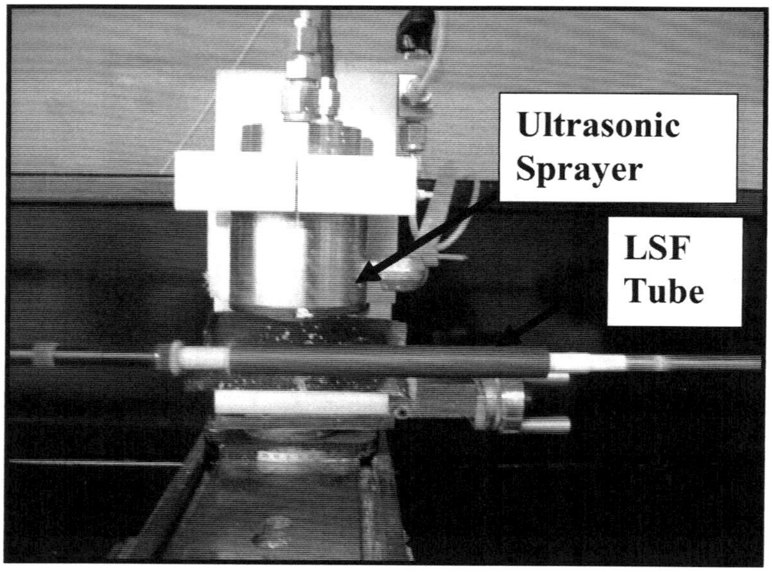

Figure 8: Automated ultrasonic spray deposition apparatus for depositing the active layers onto the tubular LSF porous support.

The layers shown in Figure 2 were deposited in multiple steps (deposition and drying steps). The LSF/GDC interlayer was sintered during the same thermal step as the GDC layer. So far, the outer LSF/GDC layer was deposited and sintered in a separate step, so that the surface of the GDC layer could be carefully characterized for cracks or micro-pores. The quality of the GDC electrolyte layer is the primary variable controlling quality of the COG element. Figure 9a shows the SEM micrograph of the surface of a LSF tube which was coated with the GDC suspension. The micrograph shows GDC spherical particles, that are <1 μm in size, uniformly packed together with some degree of dispersed porosity. The micrograph is consistent with the entire sample surface, where the surface does not show any large voids or drying cracks. Figure 9b displays the surface of a similarly GDC-coated LSF tube after thermal co-sintering stage. The micrograph shows that after co-sintering, the GDC layer is highly dense (>95%) with critical pores or cracks. This micrograph exemplifies the process when the LSF tube was properly pre-sintered, so that its dimensional shrinkage during the co-sintering process matches the LSF/GDC interlayer and GDC coating.

Figure 10 shows the secondary and backscatter SEM micrographs of a cross-section of a porous LSF tube coated and co-sintered with a LSF/GDC interlayer and GDC electrolyte layer. The coated tubular sample was co-sintered at 1400°C. The LSF/GDC outer electrode was not deposited on this sample, so this layer is not shown in the micrograph.

The micrograph shows a fairly dense GDC coating (the light colored phase in the backscatter micrograph) with a few dispersed pores. The thickness of the GDC electrolyte layer is approximately 15 μm. The pores shown in the micrograph do not span the thickness of the coating which would allow unwanted gases to penetrate through the membrane. Porosity and defect formation in the GDC electrolyte layer limits the usefulness of the COG element. In order to test the "gas-tightness" of the electrolyte layer, a vacuum was pulled on a section of the coated tube. The sample was able to hold a vacuum of 15-20 inches·Hg for 45 min (the test was discontinued after this time). This simple test allows the GDC coating of the co-sintered samples to be quickly characterized and quantified.

The LSF/GDC interlayer shown in Figure 10 (the mixed light and dark colored phase in the backscatter micrograph) that lies between the LSF porous tube and the GDC electrolyte layer. The interlayer in this micrograph is approximately 5 μm in thickness. A desired interlayer should possess an adequate amount of porosity, so that the gasses can penetrate and access the increased electrochemically active triple-phase boundary area. The interlayer displays a porosity level <20%. Therefore, further research will focus on increasing the thickness of and amount of micro-porosity in the interlayer. The same work will be applied to the deposition and co-sintering of the outer LSF/GDC electrode. going work at NexTech is focused on further refining this process and integrating the composite interlayer into the spray deposition processes.

Figure 9. SEM micrographs of the a) green and b) sintered GDC coating on the surface of the LSF tubular supports.

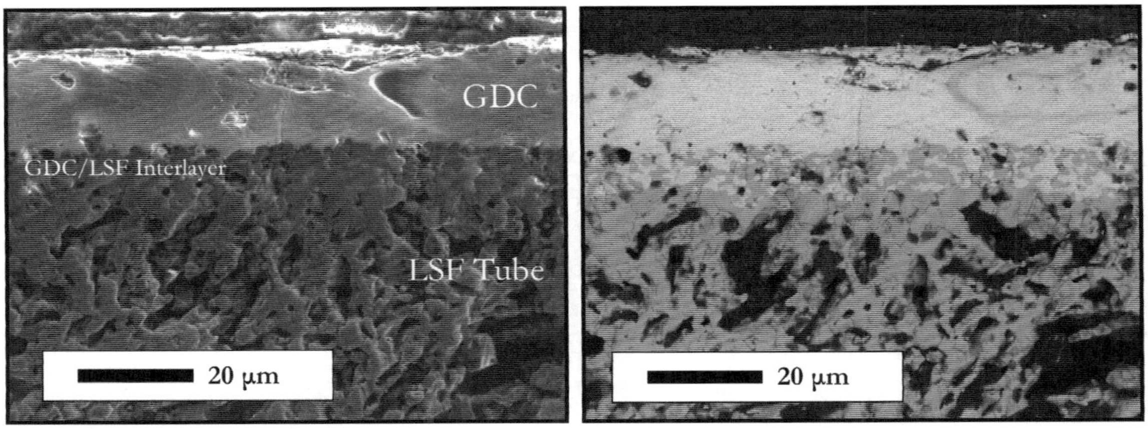

Figure 10. Secondary and backscatter SEM images of the cross-sectional view of a LSF tube coated and co-sintered with a LSF/GDC interlayer and GDC electrolyte layer.

The previous micrographs show the flexibility of aerosol spray deposition process. The porosity, chemistry and phase distribution of each active layer can be effectively modified and controlled by varying the spray suspension, the thermal schedule, and a few spray parameters. Figure 11 shows a picture of two COG elements after co-sintering interlayer and GDC electrolyte layers. At this point, a similar spray suspension used for the interlayer will be applied to the outer surface of these elements and sintered, leaving a highly active mixed-conductive porous electrode layer. Currently, NexTech is now beginning to test the oxygen permeability of the COG elements by measuring the conductivity through the tubes under a DC-bias in air. Also, a single element prototype reactor is being fabricated in order to quantify the oxygen permeation of COG elements at various electrical drives and reactor pressures.

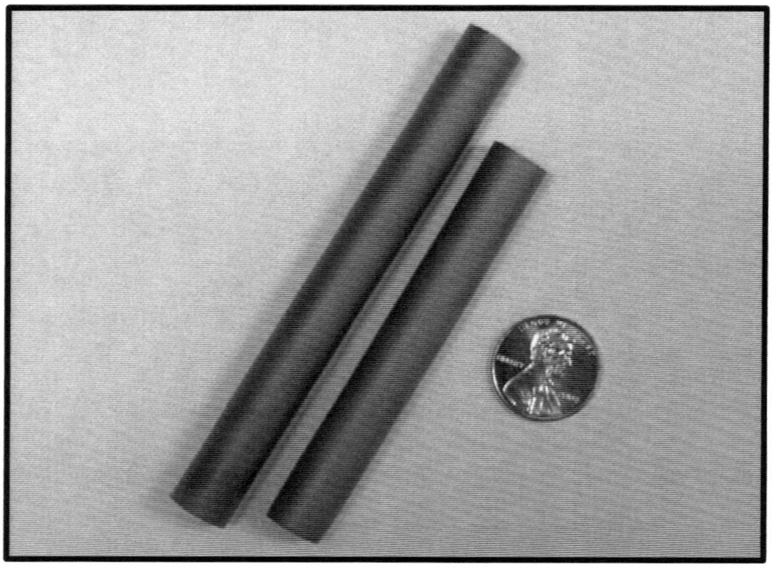

Figure 11. Picture of COG elements fabricated by aerosol spray depositing the active layers and co-sintering the layers on a partially sintered LSF porous tube.

Conclusions and Future Work:

Processes for synthesizing doped-CeO_2, LSF, and doped-LSF powders, which all provide high electrical performance at low temperatures, were developed. These powders were utilized in the development of a cost-effective and efficient manufacturing process for fabricating tubular electrode-supported, thin-film ceramic oxygen generation (COG) elements. In order to produce the COG elements, the following ceramic processing methods were identified and developed during this program: 1) extrusion and sintering processes for fabricating porous LSF electrode tubes, 2) aerosol-spray suspensions and deposition processes for coating the LSF electrode tubes, and 3) co-sintering processes for matching the shrinkage of the LSF electrode tube, LSF/GDC interlayer and GDC electrolyte layer, while still maintaining the porosity of the electrode tube as well as producing a dense GDC membrane. The fabrication experiments completed through this program have demonstrated that the aerosol-spraying and co-sintering processes offer a very versatile process for depositing multiple ceramic thin films with different chemistry and microstructure.

Future work for this program will focus on the incorporation of the doped-LSF/GDC composite interlayer compositions between the LSF tubular support and the outer electrode and the GDC electrolyte layer. Also, the quality of the COG elements will be measured by current/voltage measurements (DC-bias) and/or prototype reactor experiments. These experiments will be completed in order to quantify the effect of the microstructure and electrode interlayer compositions on the overall cell resistance and oxygen output.

Acknowledgements:
This work has been supported by they Department of Energy Office of Industrial Technology, Contract Number DE-FG02-01ER832251. Electrode development work has also been supported by the Department of Energy Fossil Energy Program, Contract Number DE-FG-02-01ER83227

22c Novel Polymer Nanocoating Processing to Produce Ceramic Nanocomposites

Alan Weimer, John Ferguson, Joseph Spencer, Karen J Buechler, Steven George

High density polyethylene (HDPE) particles having an average particle size of 66 microns are fluidized and coated by Atomic Layer Deposition (ALD) with 1 to 10 nanometer thick films of alumina. The sequential precursors used to grow the alumina were trimethyl aluminum (TMA) and water vapor. The processing was carried out at 77 degrees Celsius and 625 millitorr pressure with a fluidized bed of approximately 100 grams of HDPE. A physical vibrational frequency of 3600 Hz was applied to the bed to aid in the fluidization.

The coated particles were characterized by transmission electron microscopy (TEM) and scanning electron microscopy (SEM), inductively couple plasma – atomic emission spectroscopy (ICP-AES), surface area, and particle size distribution. It was determined that individual HDPE particles were nanocoated rather than aggregates of particles. The coated polymer particles were extruded using a Bonnot heated BB-gun extruder. The extruded materials were characterized by cross sectional TEM imaging. Alumina shell remnants were distributed throughout the polymer matrix.

The ability to nanocoat individual HDPE particles with ceramic films provides for an intimate mixture of polymer and nano length scale ceramic material prior to any mixing operations. Hence, the slow kinetics of blending nanothick ceramic platelets into a polymer matrix is overcome. It is possible to vary the ceramic layer composition and thickness to achieve particular properties for barrier or impact resistance.

CMS

Novel Air/VOC Separation Process

J. Bowser, S. Nemser*
Compact Membrane Systems, Inc.

R. Grantham
Vapor Systems Technologies, Inc.

*Presenting

Presentation

- Motivation/Goals
- Proposed System
- Some Learnings from the Development
- Field Performance

AIChE April 2005

Motivation for This Work

Gasoline service stations operating without vapor controls lose between one and five gallons of gasoline to the air for every 1000 gallons pumped.

- These VOC's contribute to the formation of smog, a recognized health hazard.
- Non-renewable energy in the form of gasoline is lost, and the station has an unnecessary cost.

Program Goal

Develop a membrane based vapor processor that will maintain fugitive emissions from retail gasoline stations below the new California Air Resources Board limit of 0.38 lb of gasoline lost per 1000 gallons of station throughput.

CMS

Vapor Recovery at Service Stations

Phase I

Diagram courtesy of California Air Resources Board

Phase II

AIChE April 2005

ORVR - Phase II Incompatibility

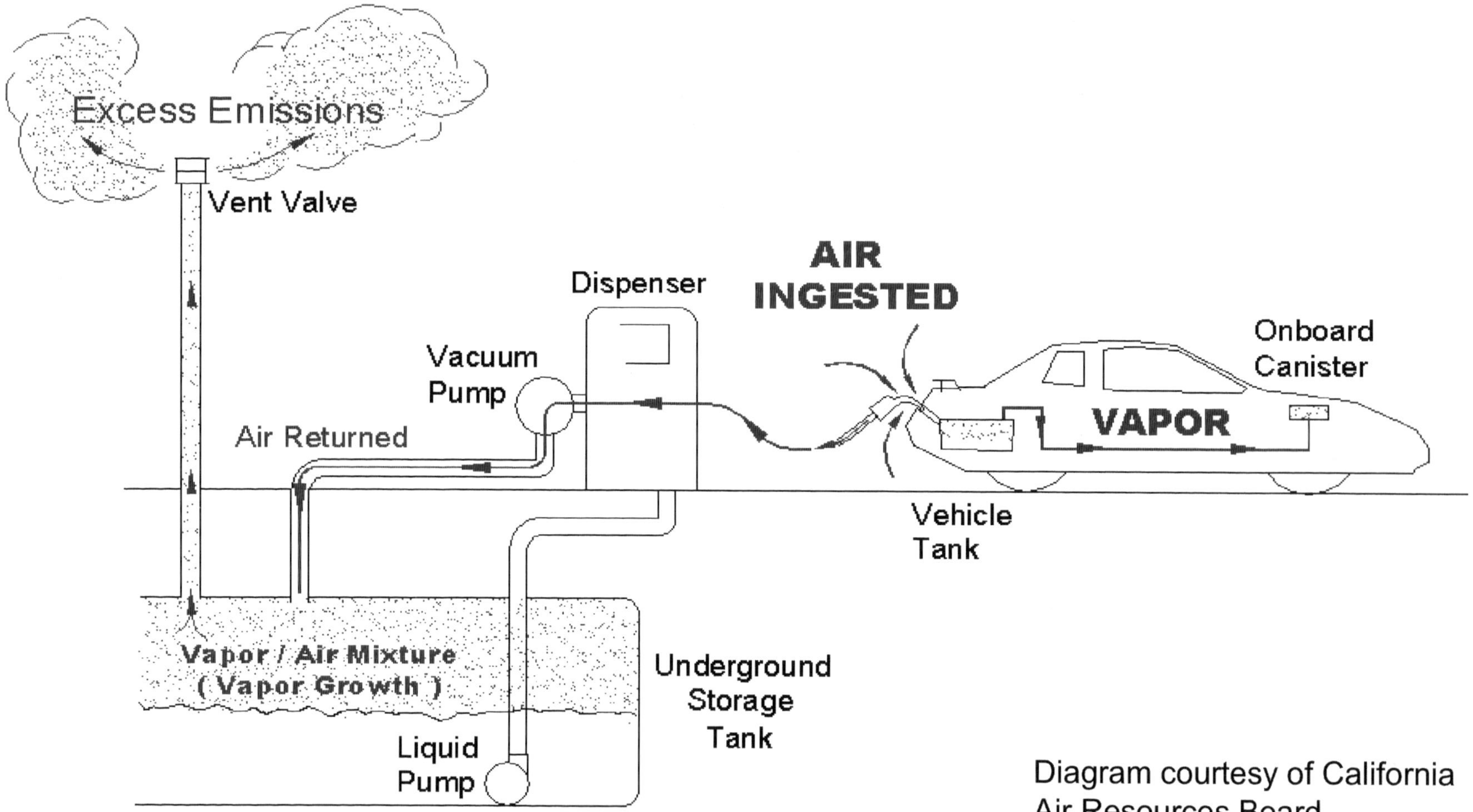

Diagram courtesy of California Air Resources Board

AIChE April 2005

Onboard Refueling Vapor Recovery or "ORVR"

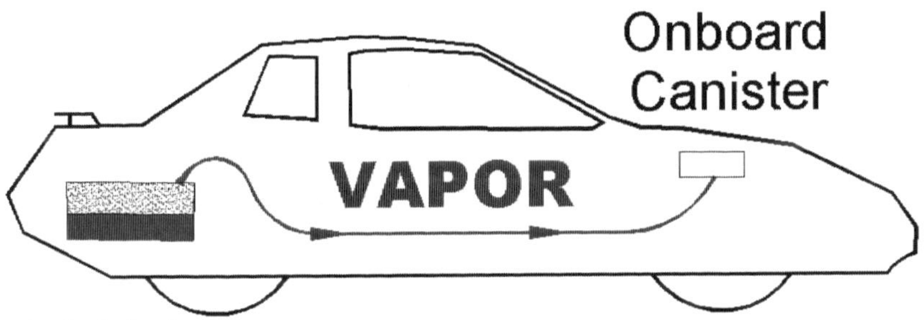

Diagram courtesy of California Air Resources Board

Vehicle Class	40%	80%	100%
Passenger	1998	1999	2000
LD Trucks & MDV (<6000 lbs)	2001	2002	2003
MD Vehicles (6001-8500 lbs)	2004	2005	2006

CMS

ORVR Penetration Projection

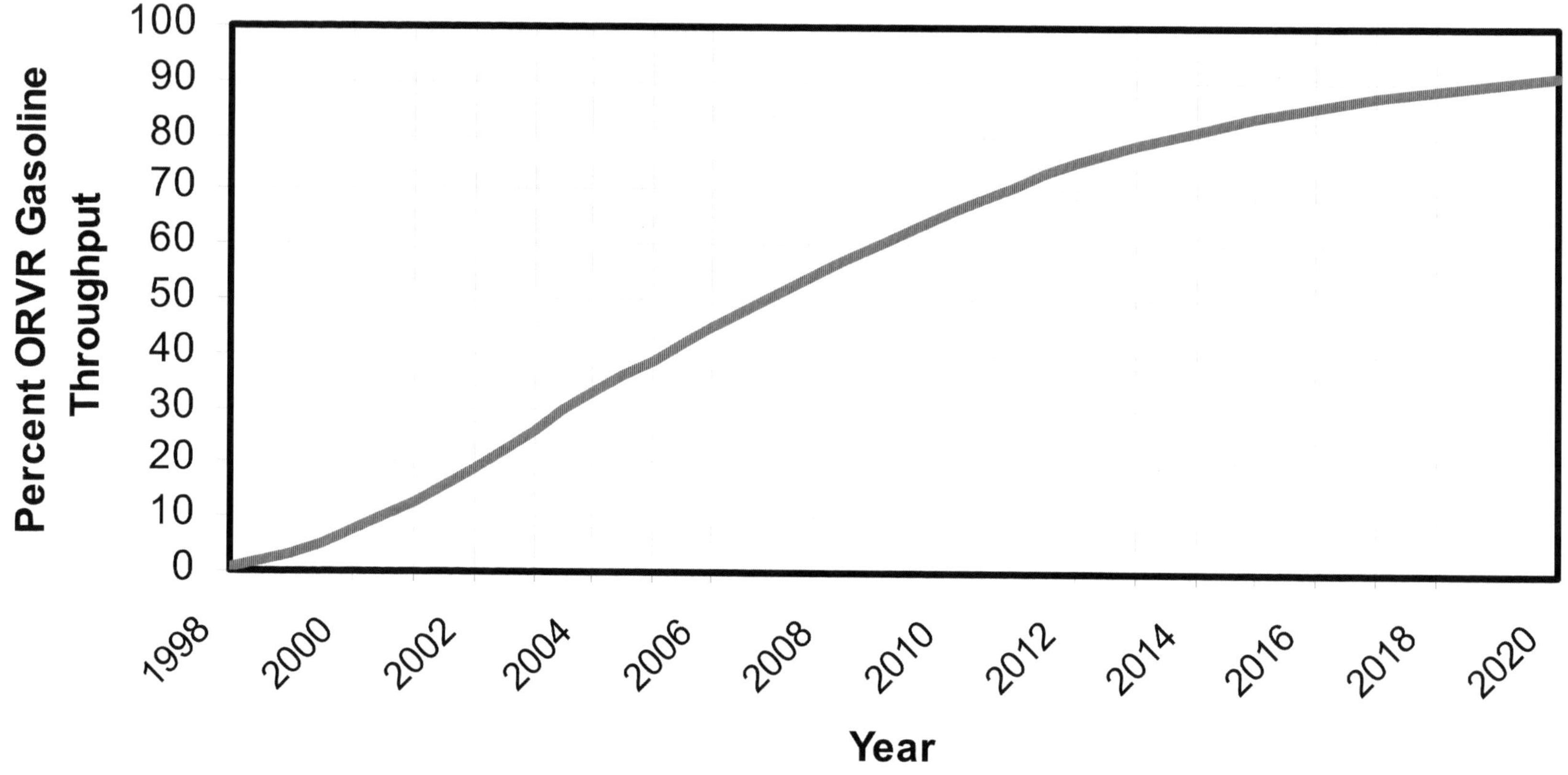

Chart courtesy of California Air Resources Board

AIChE April 2005

Vapor Recovery Provides Huge Benefits

VOC Reduction, 20 Tons/day Per division

CA South Coast Air Basin

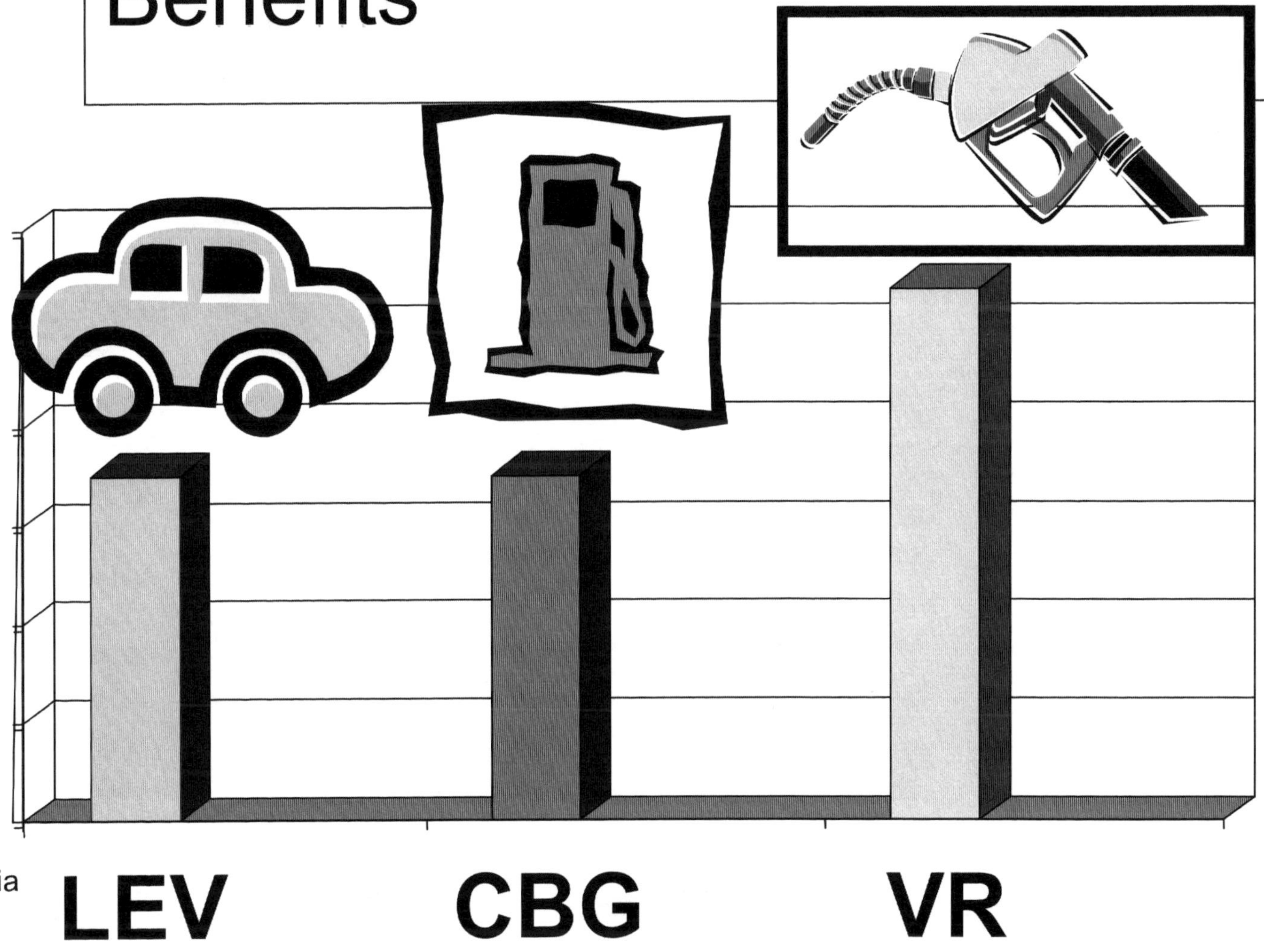

Chart courtesy of California Air Resources Board

AIChE April 2005

Vapor Processor Design

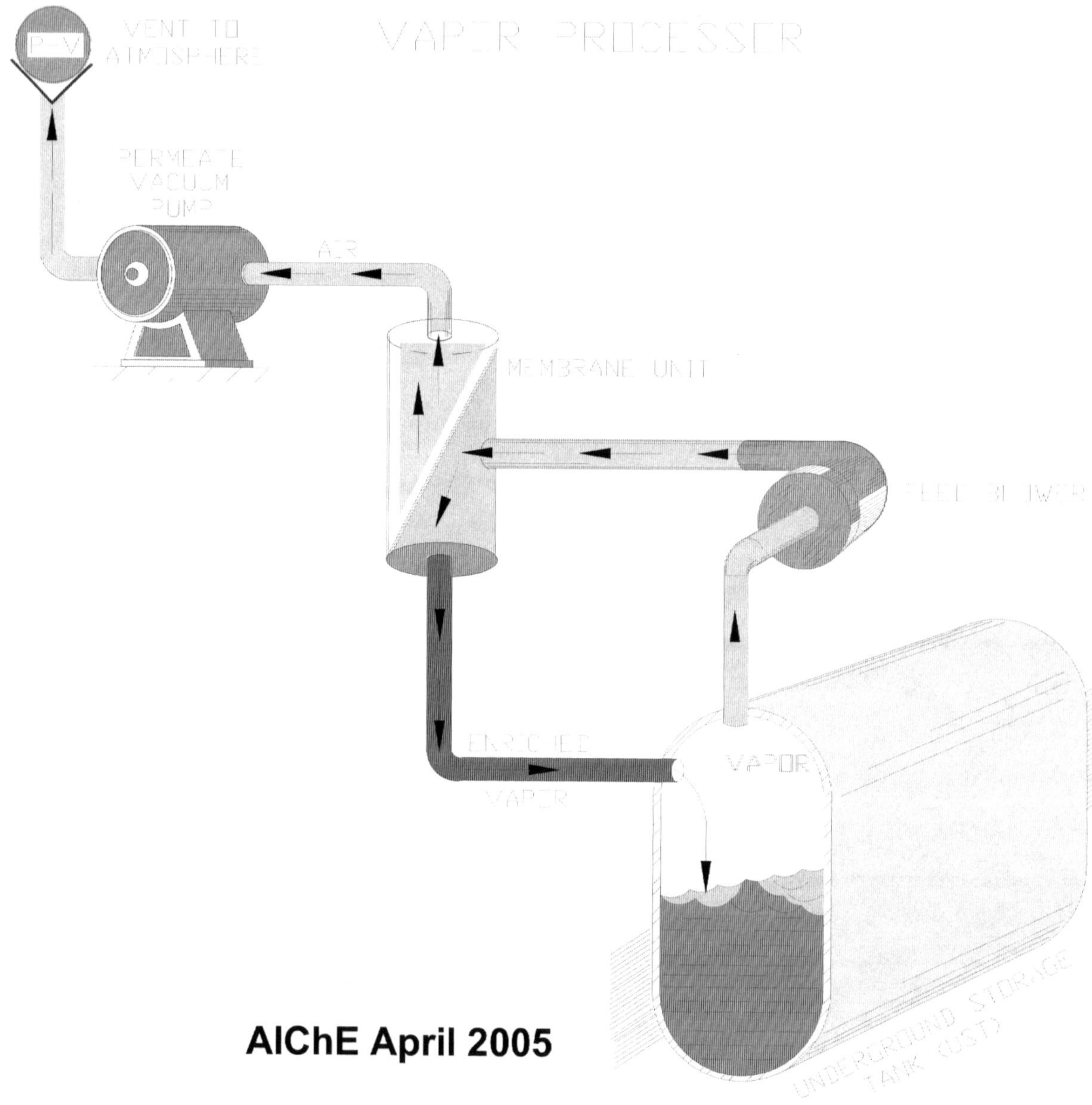

AIChE April 2005

CMS

Vapor Processor

Membrane Module

Permeate Pump

Feed Blower

AIChE April 2005

Composite Hollow Fiber

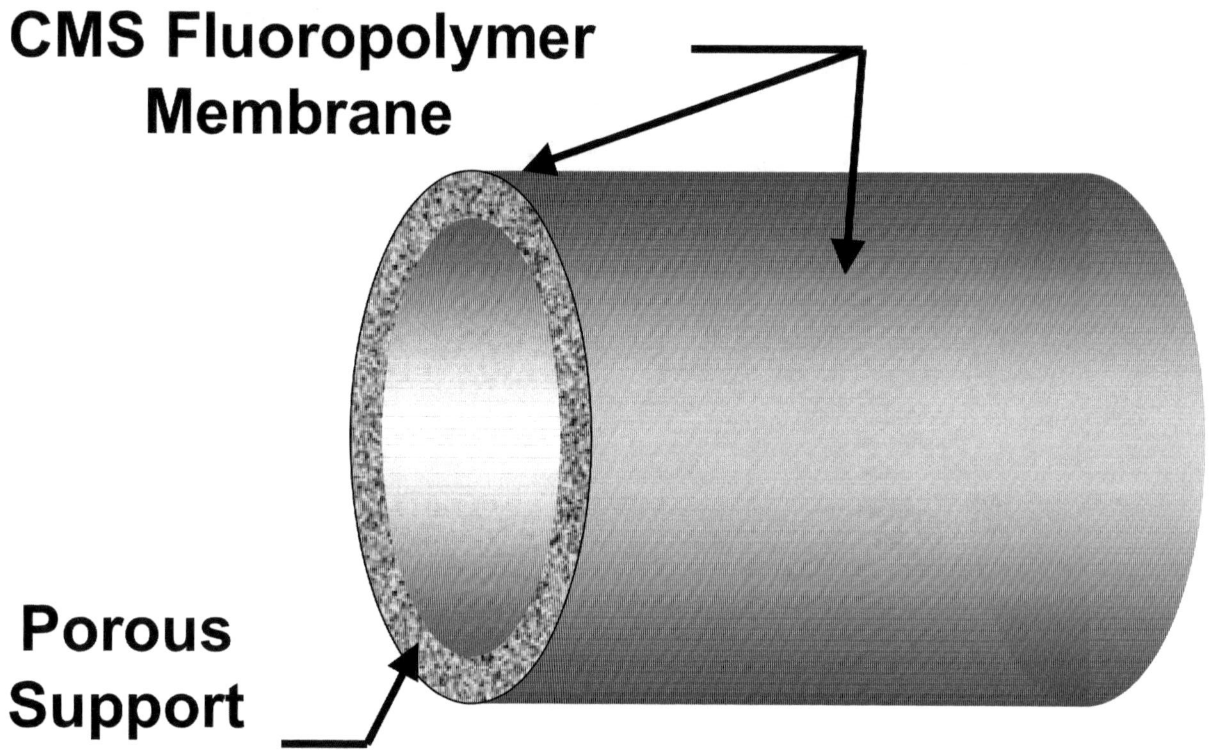

AIChE April 2005

Permeability Decreases With Molecular Size

The ratio of perms for any two gases is a good first estimate of the separation factor.

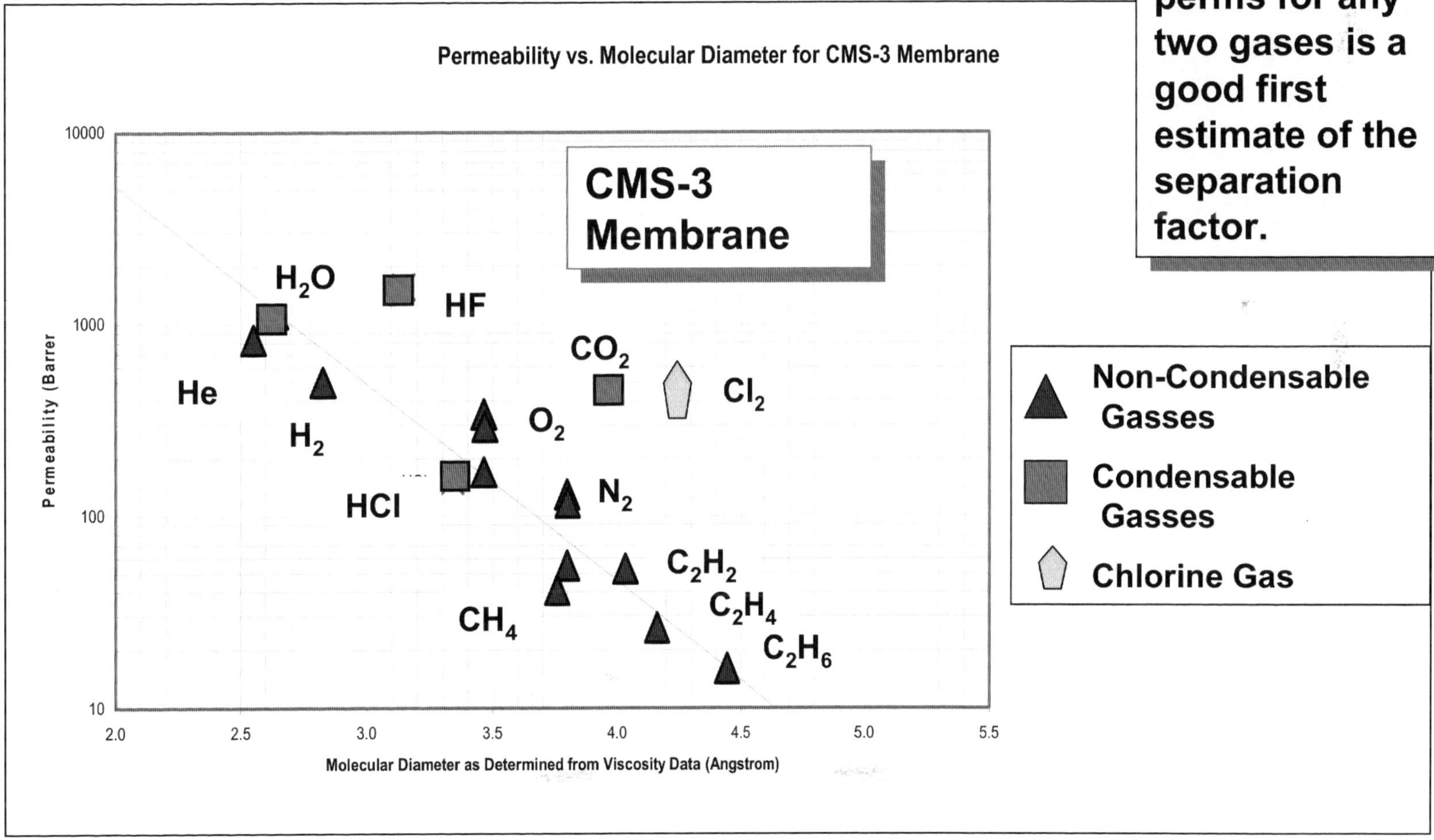

AIChE April 2005

Gas Transport in Polymers

Feed Pressure — Permeate Pressure

$P_{feed,i}$ > $p_{perm,i}$

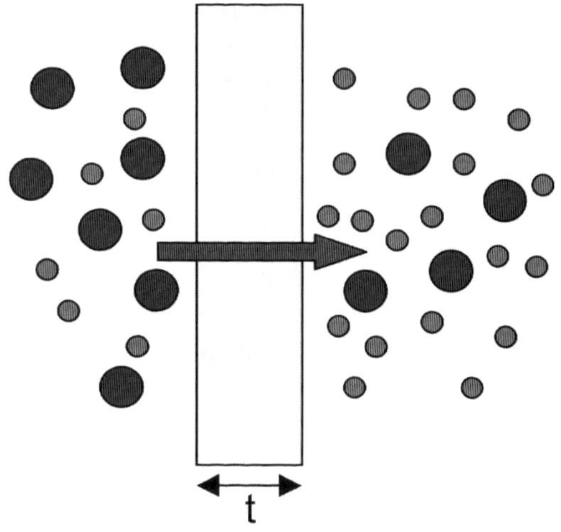

- Component A
- Component B

- Permeation Equation:

$$R_A = \frac{P_A\, A\, (P_{feed,A} - p_{perm,A})}{t}$$

- Permeability of A, $P_A = D_A\, S_A$

  Where, D_A = Diffusion Coefficient of A
  S_A = Solubility Coefficient of A

- Selectivity

$$\alpha_{A\text{-}B} = \frac{P_A}{P_B} = \left(\frac{D_A}{D_B}\right)\left(\frac{S_A}{S_B}\right)$$

Mobility Selectivity — Solubility Selectivity

AIChE April 2005

CMS

Permeability Definition

$$\text{Permeability} = \frac{\text{Volume} \cdot \text{Thickness}}{\text{Area} \cdot \text{Time} \cdot \Delta\,\text{Pressure}}$$

$$[\text{barrers}] = 10^{-10}\ \frac{\text{cm}^3\,(\text{STP}) \cdot \text{cm}}{\text{cm}^2 \cdot \text{sec} \cdot \text{cm Hg}}$$

Permeance (productivity) is higher for thinner membranes

$$\alpha_{\text{gas 1/ gas 2}} = \frac{\text{Fast Gas Permeance}}{\text{Slow Gas Permeance}}$$

AIChE April 2005

CMS-3 Amorphous Perfluoropolymer Membrane

- Glassy polymer – passes atmospheric gasses and retains hydrocarbons above C_2
- Highest permeance commercially available for atmospheric gasses – small modules and associated equipment
- Low affinity for hydrocarbons – non-fouling, maintains flux over time
- Perfluorinated – Super chemical and temperature resistance

AIChE April 2005

Membrane Cartridge

AIChE April 2005

Prevents Fugitive Emissions by Holding Pressure in the UST Below Atmosphere

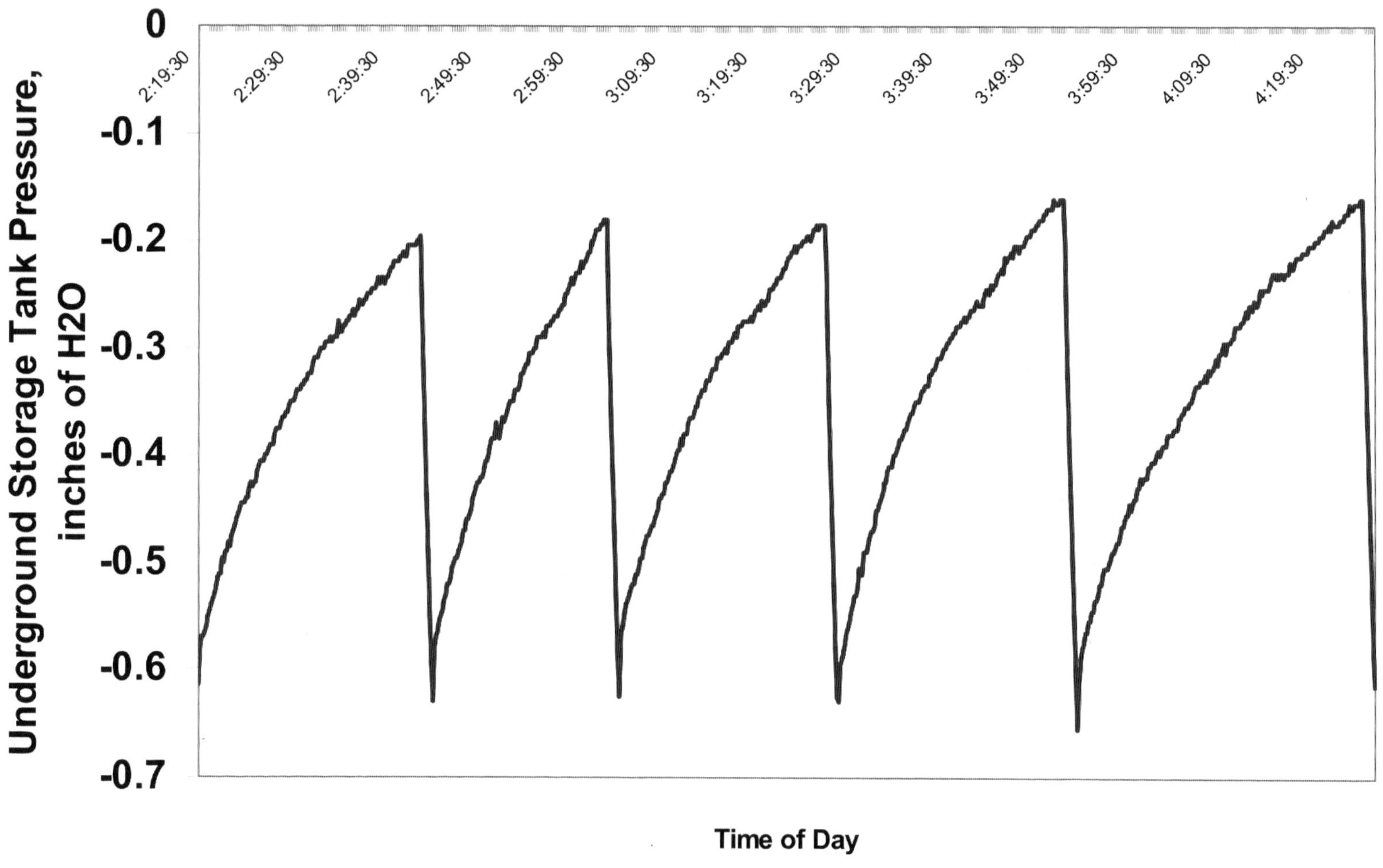

AIChE April 2005

UST Pressure over One Day with Vapor Processor Operating

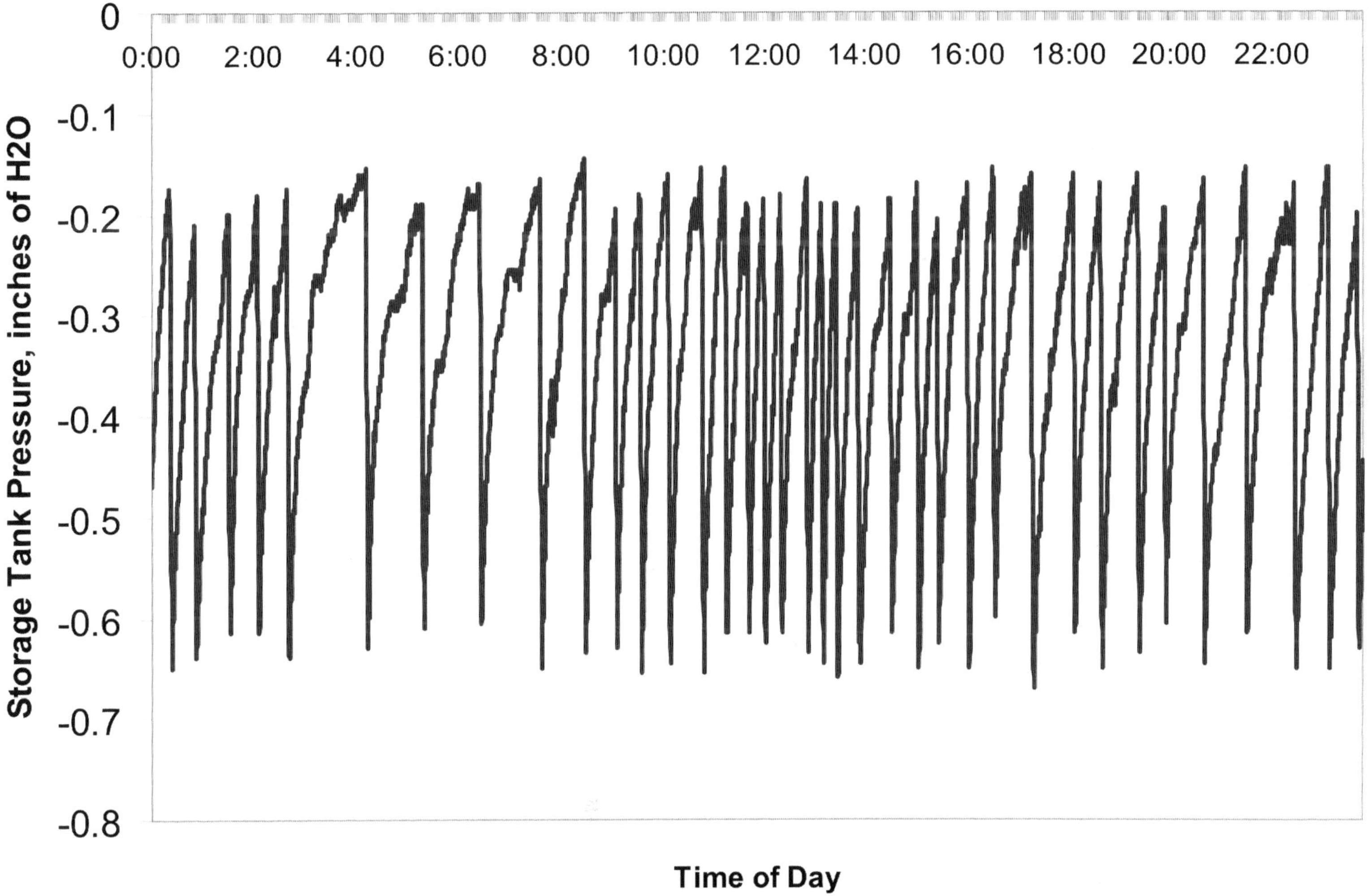

AIChE April 2005

Winter Fuel Vapor Components

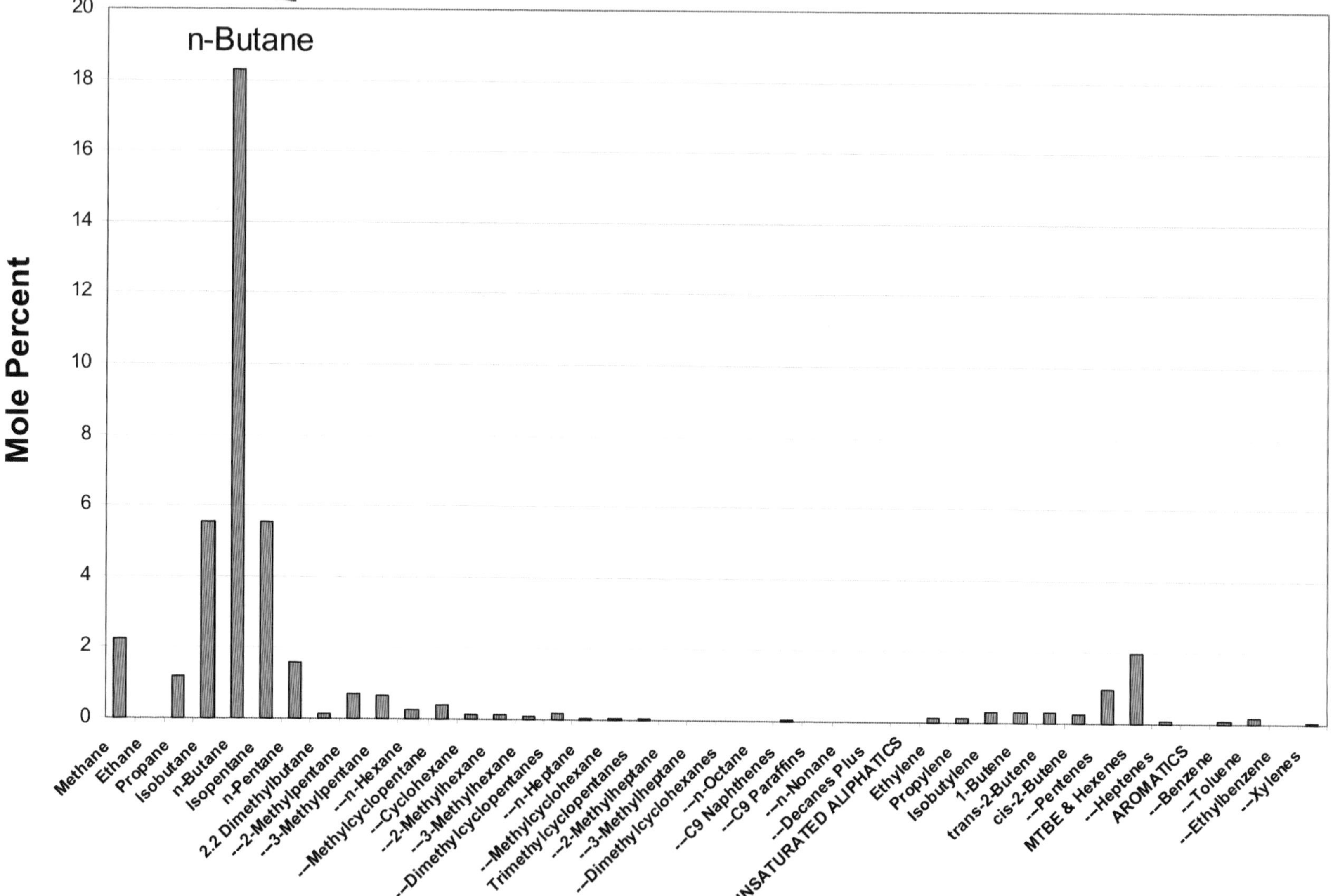

AIChE April 2005

Summer Fuel Vapor Components

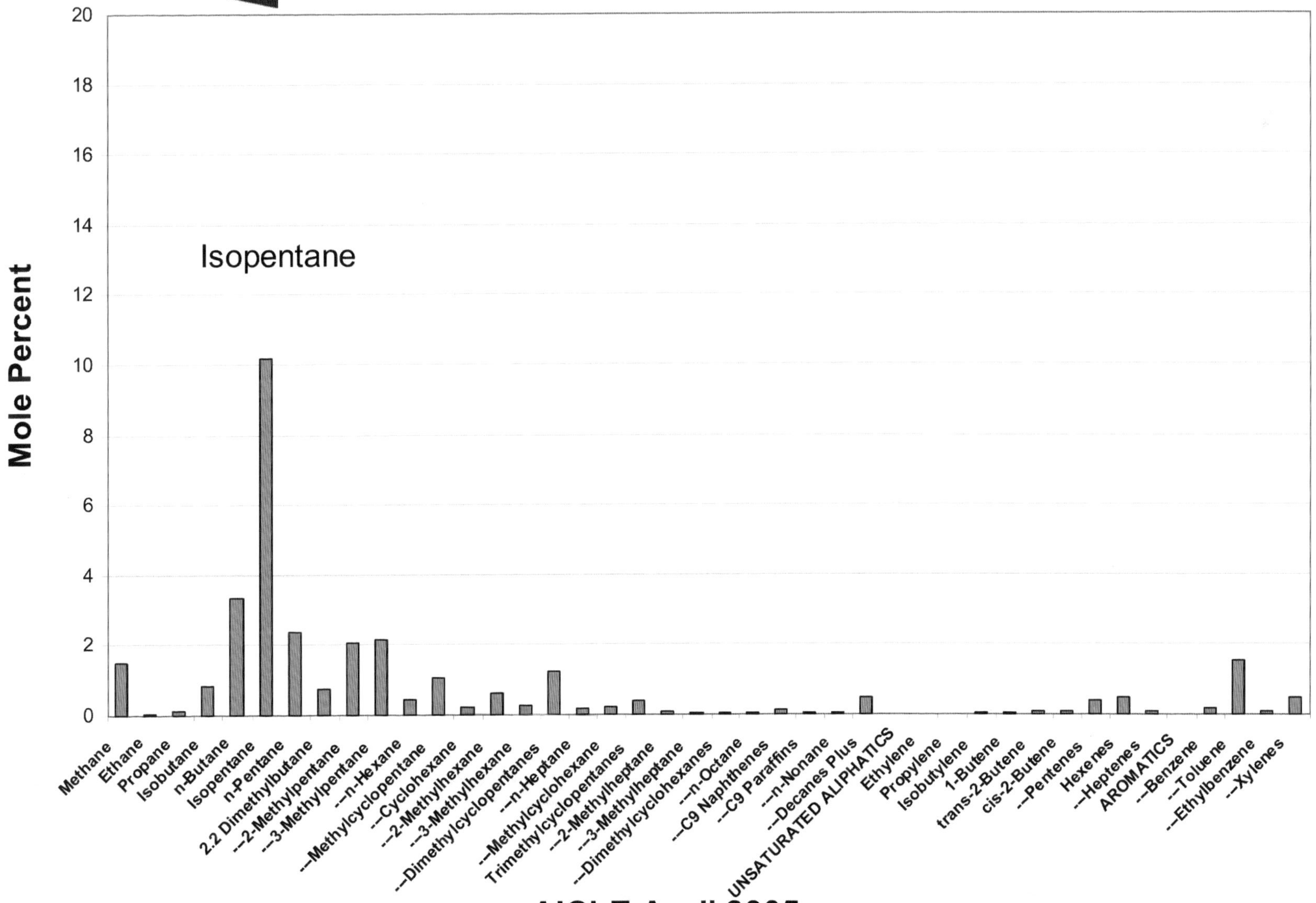

AIChE April 2005

Processor Reduction Ratio for Selected Gasoline Components

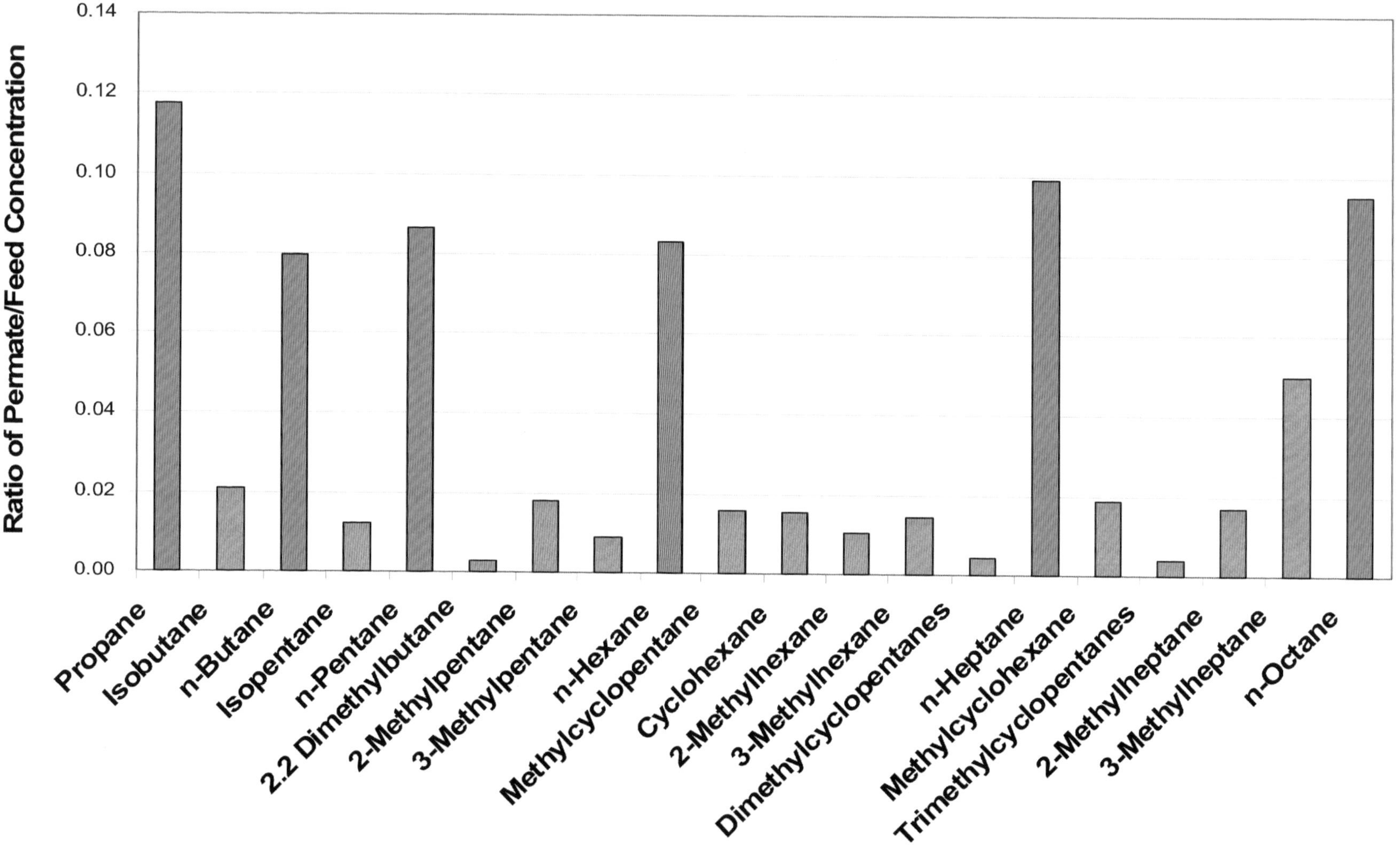

AIChE April 2005

Vented VOC Surge on Processor Startup

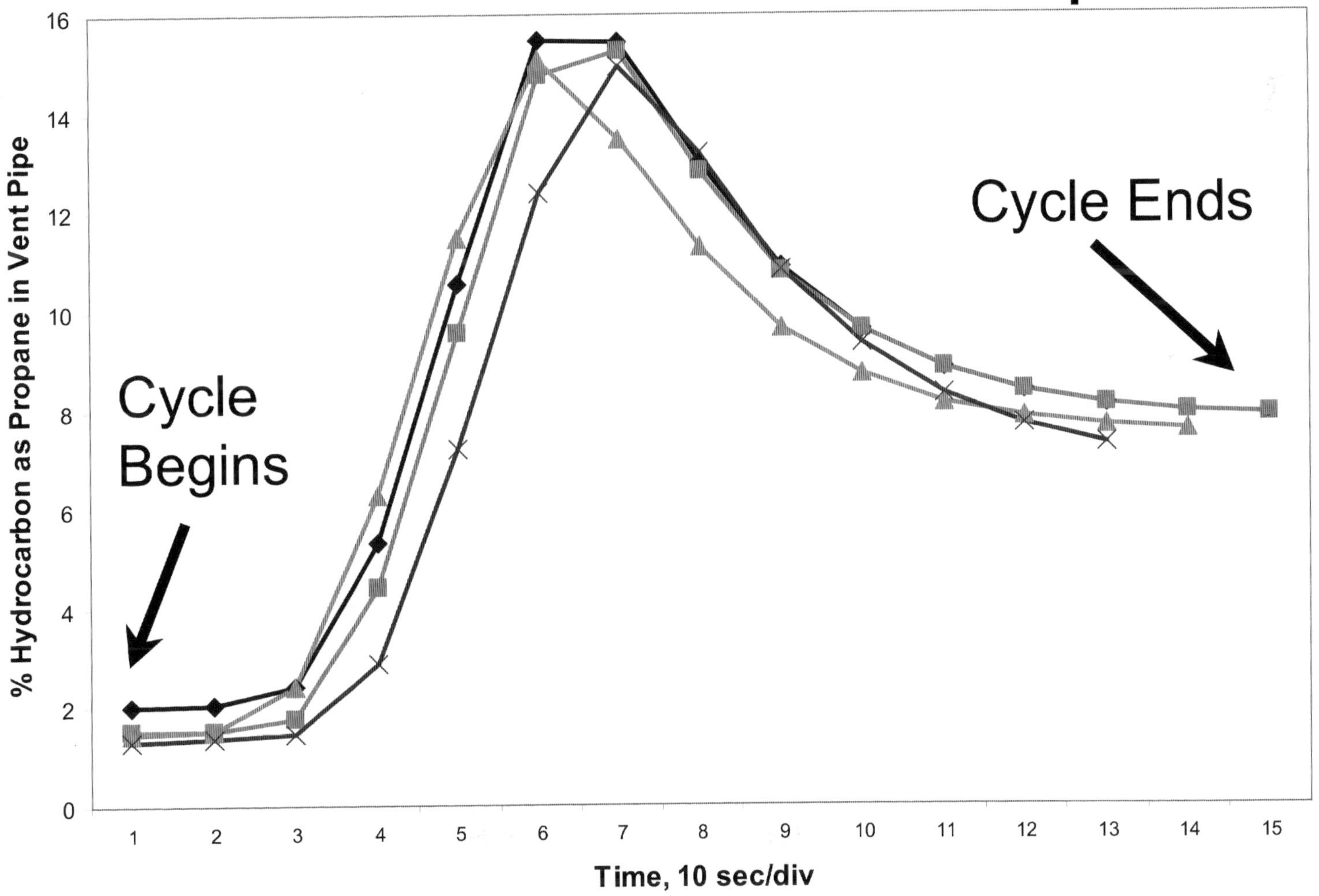

AIChE April 2005

Vapor Processor Design

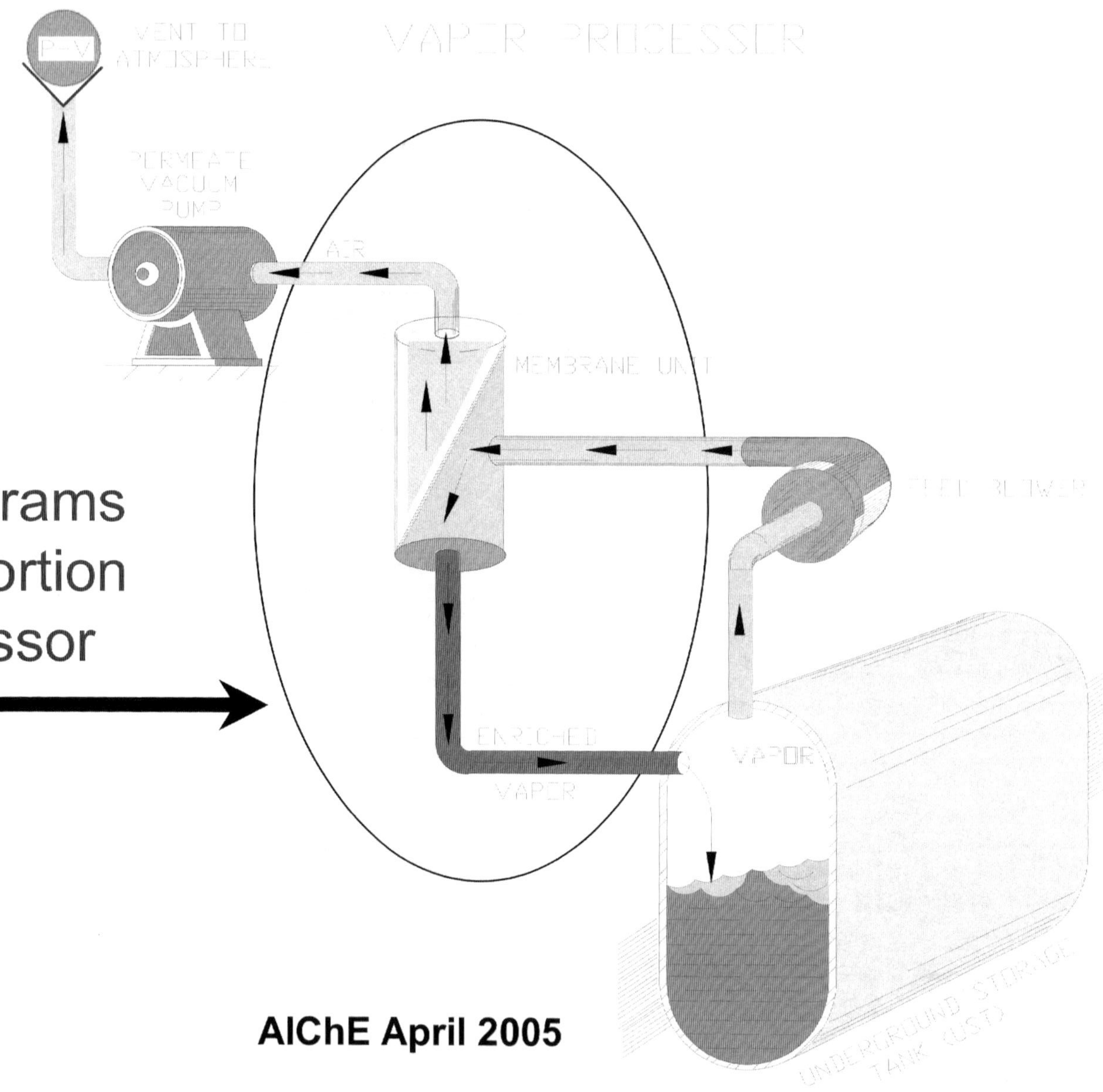

Next few diagrams refer to this portion of the processor

AIChE April 2005

Why the VOC Surge Occurs

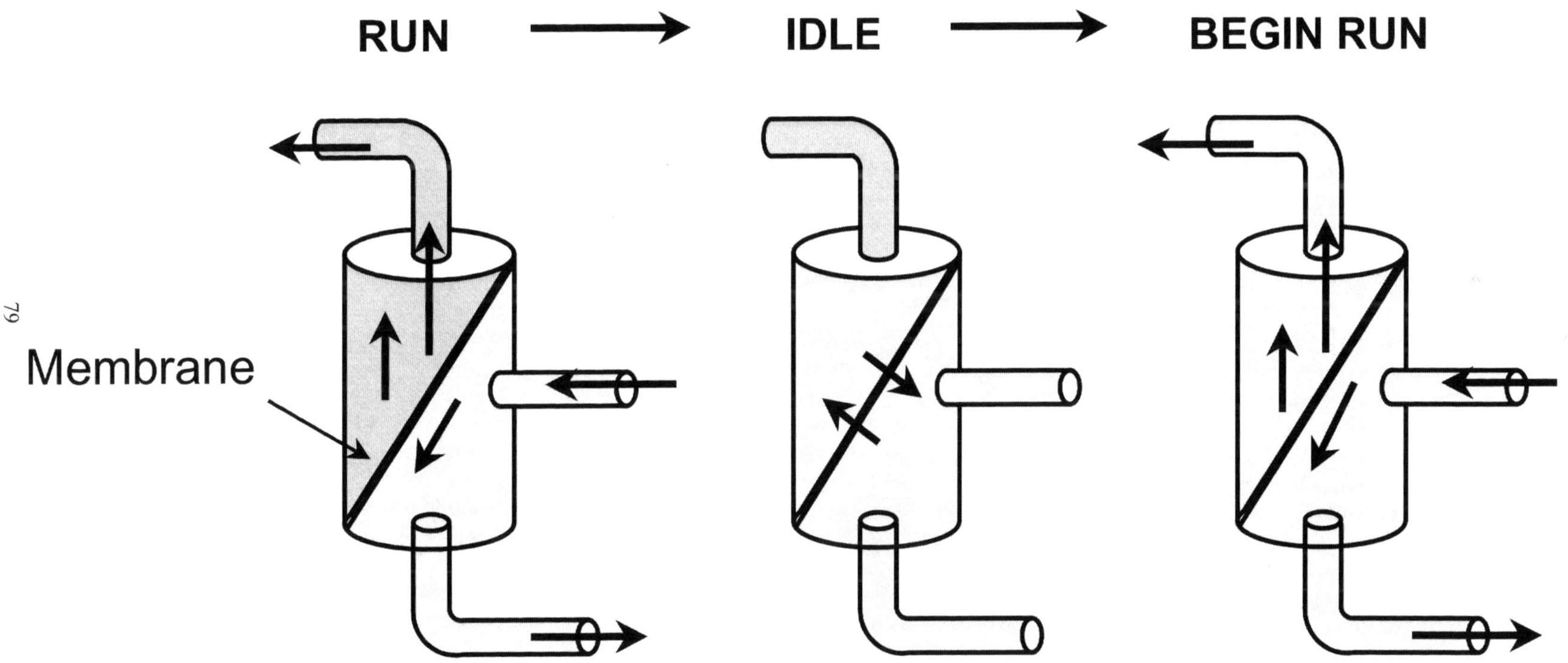

AIChE April 2005

A Remedy for the VOC Surge?
Briefly purge the module with air.

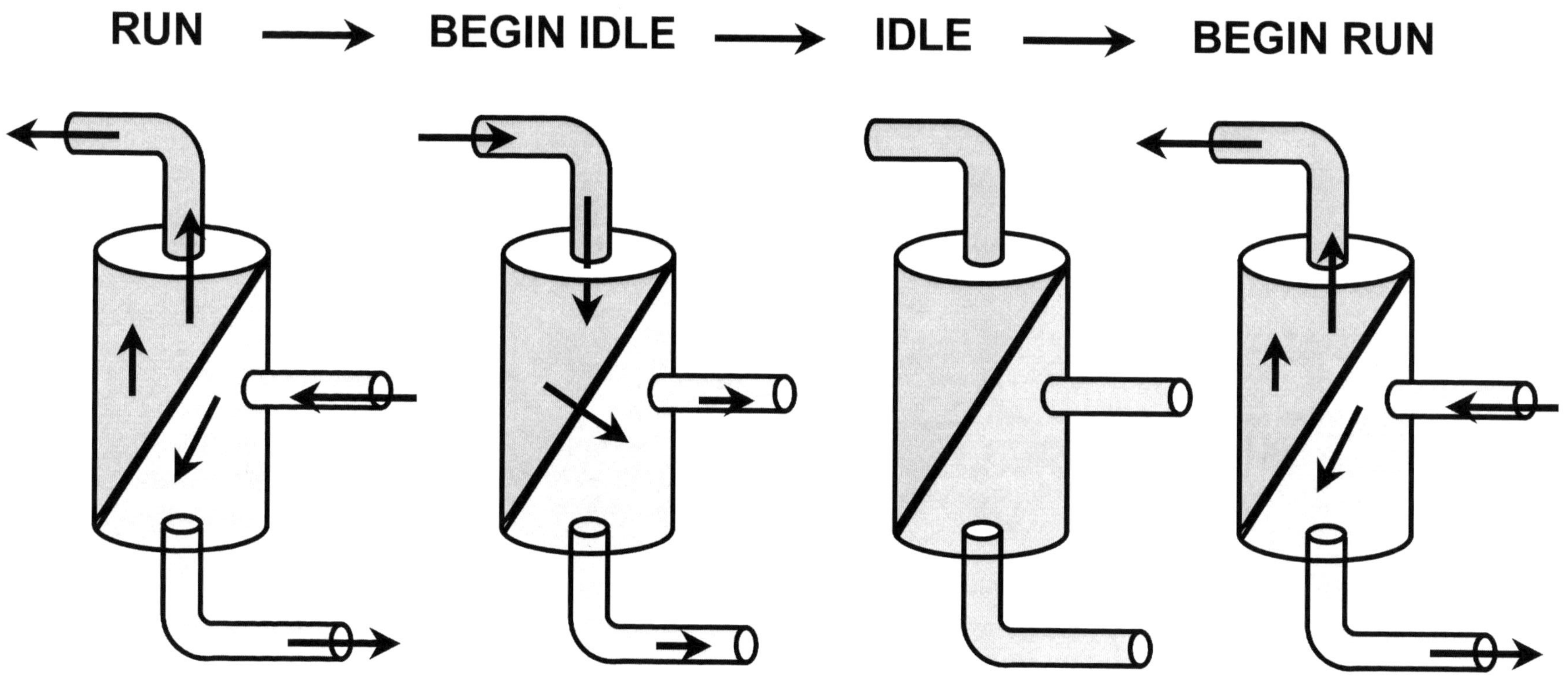

AIChE April 2005

Vented VOC Profile with Module Purge

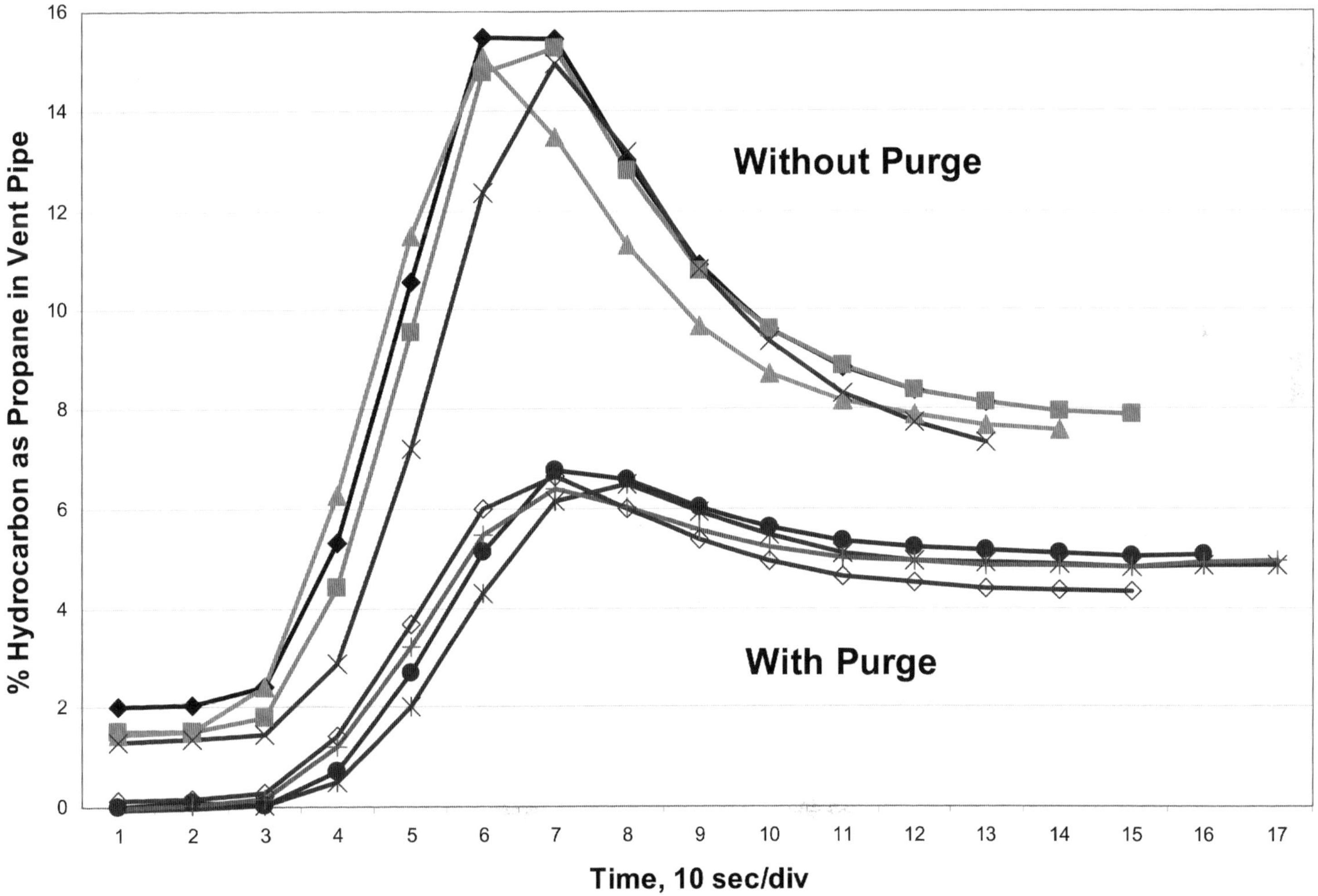

AIChE April 2005

Key Improvements to Vapor Processor

- Introduction of Module Purge reduced emissions by a factor of 2.
- Development of a composite hollow fiber membrane with significantly improved separation capability reduced emissions by a further factor of 2.

AIChE April 2005

CMS

UST Vents

Vapor Processor

AIChE April 2005

Performance

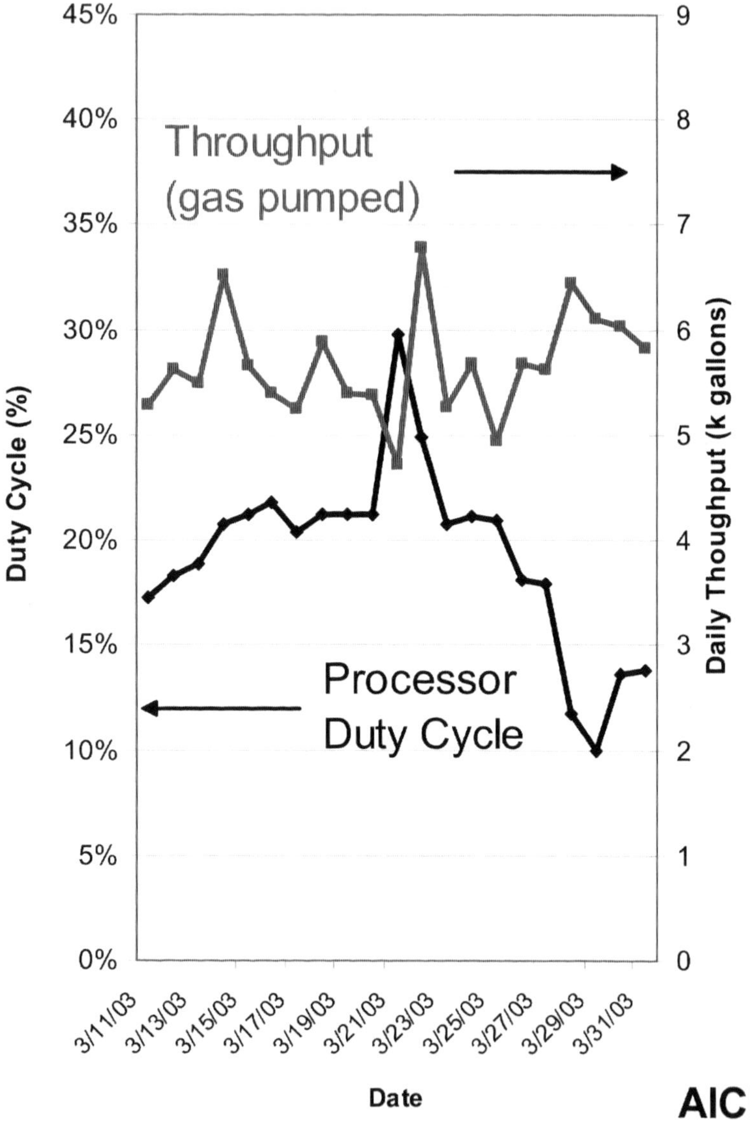

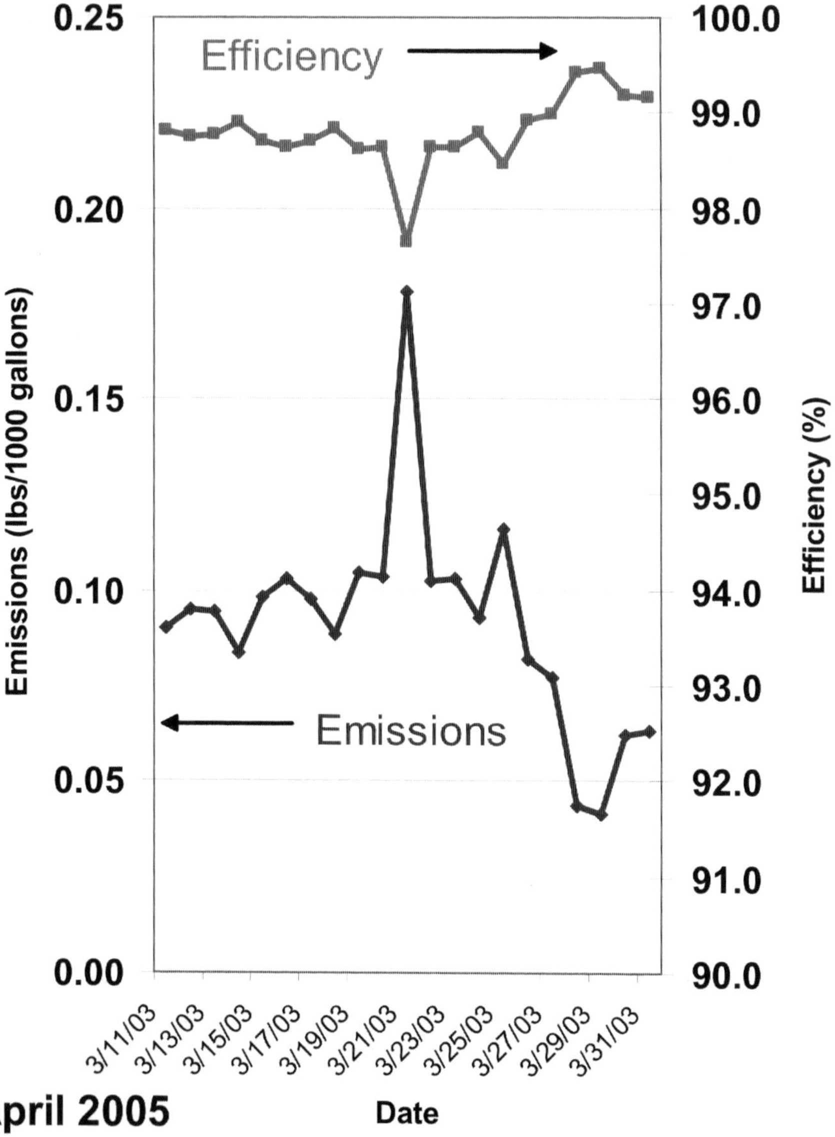

AIChE April 2005

Performance

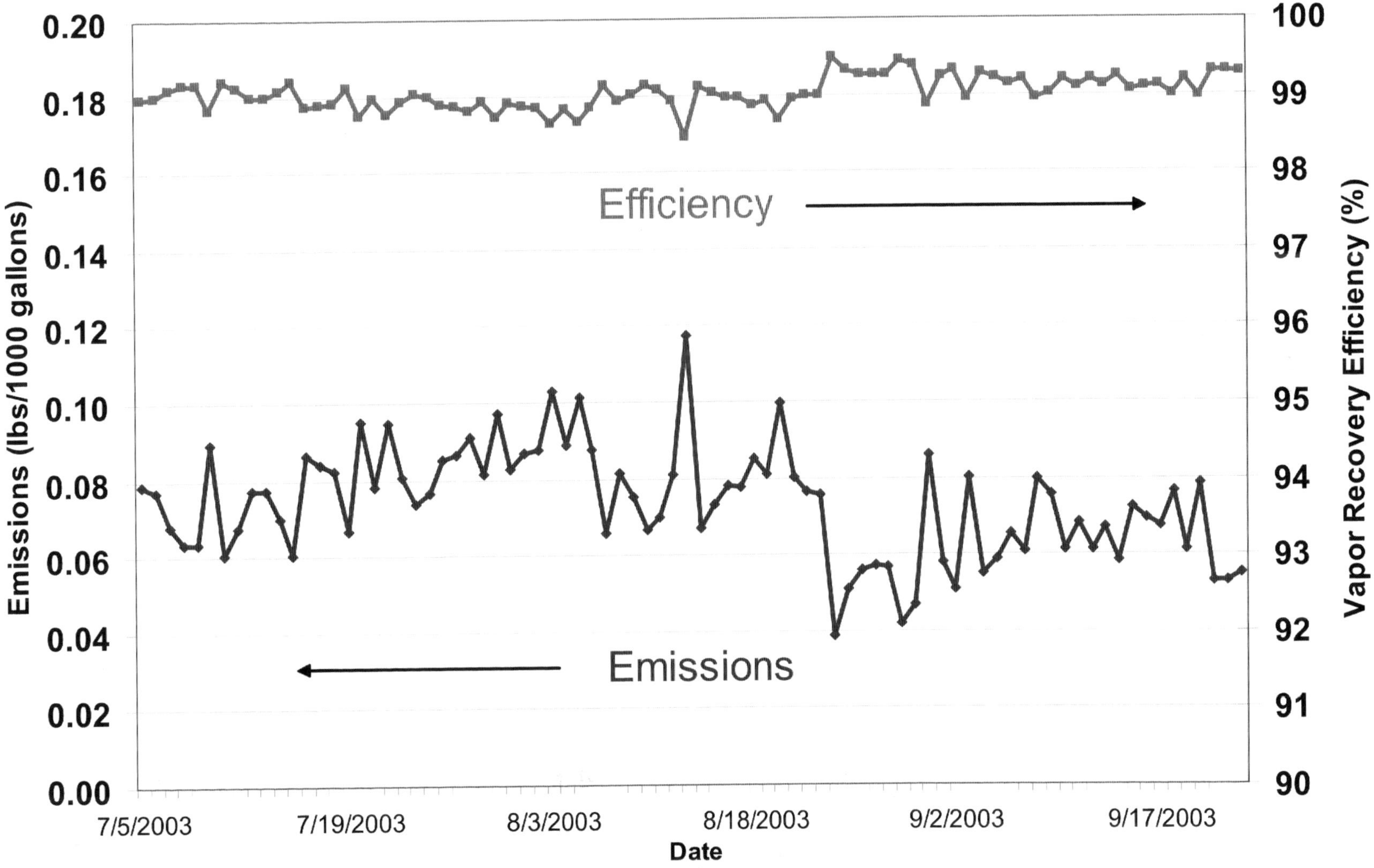

AIChE April 2005

In Conclusion

- A simple membrane based gasoline vapor recovery system was developed that easily exceeds the new CARB standards.
- We expect certification of the equipment by CARB to occur in 4Q, 2005.
- CMS will supply the membrane unit, Vapor Systems Technologies will supply the processor.

CMS

Acknowledgement

The authors gratefully acknowledge the support of the **US Department of Energy** through SBIR Contract DE-FG02-00ER82976

CMS

Extra Slides Follow

Stu, that was the end of the slide show. You can stop now.

Onboard Refueling Vapor Recovery or "ORVR"

Diagram courtesy of California Air Resources Board

Vehicle Class	40%	80%	100%
Passenger	1998	1999	2000
LD Trucks & MDV (<6000 lbs)	2001	2002	2003
MD Vehicles (6001-8500 lbs)	2004	2005	2006

ORVR - Phase II Incompatibility

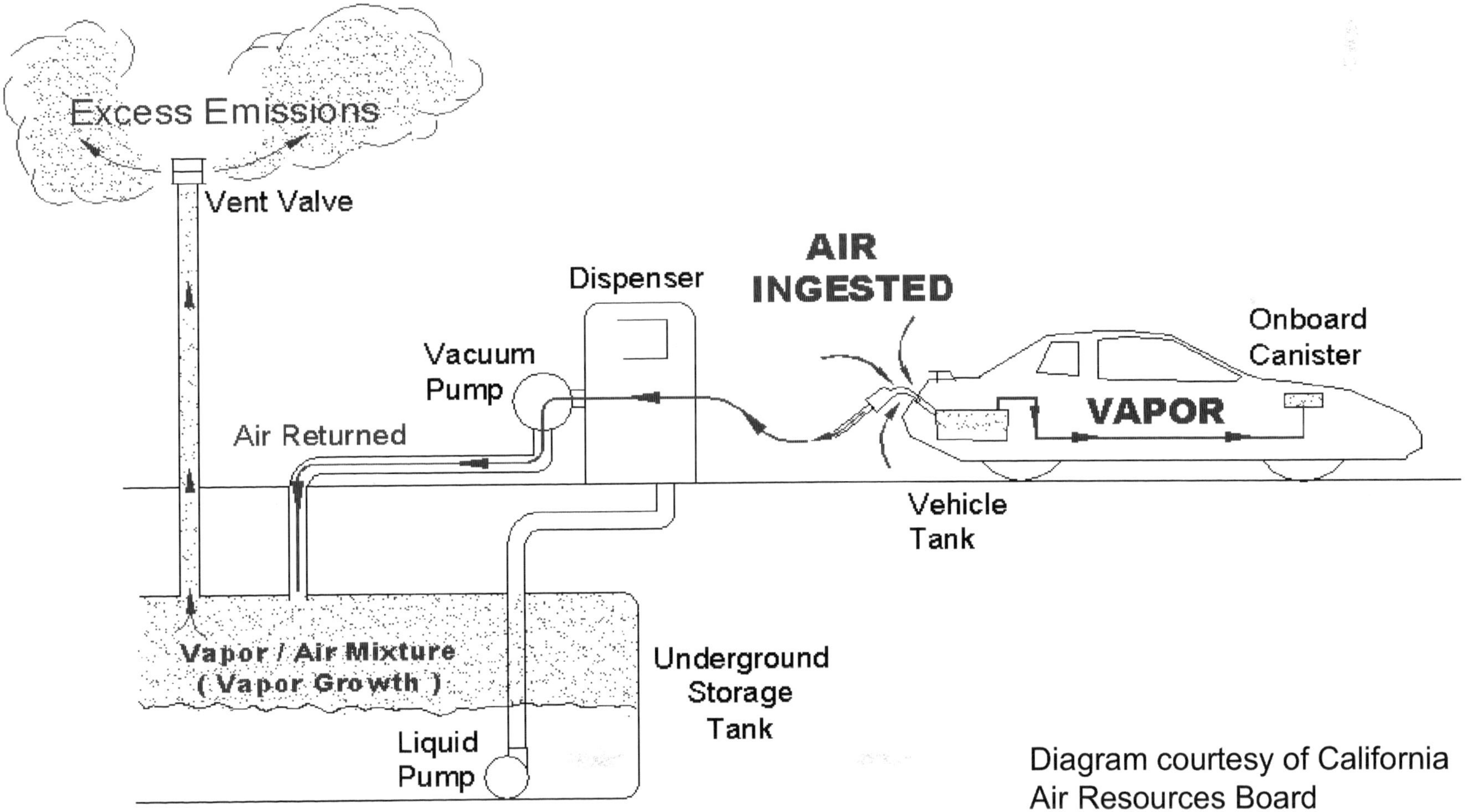

Diagram courtesy of California Air Resources Board

AIChE April 2005

ORVR Penetration Projection

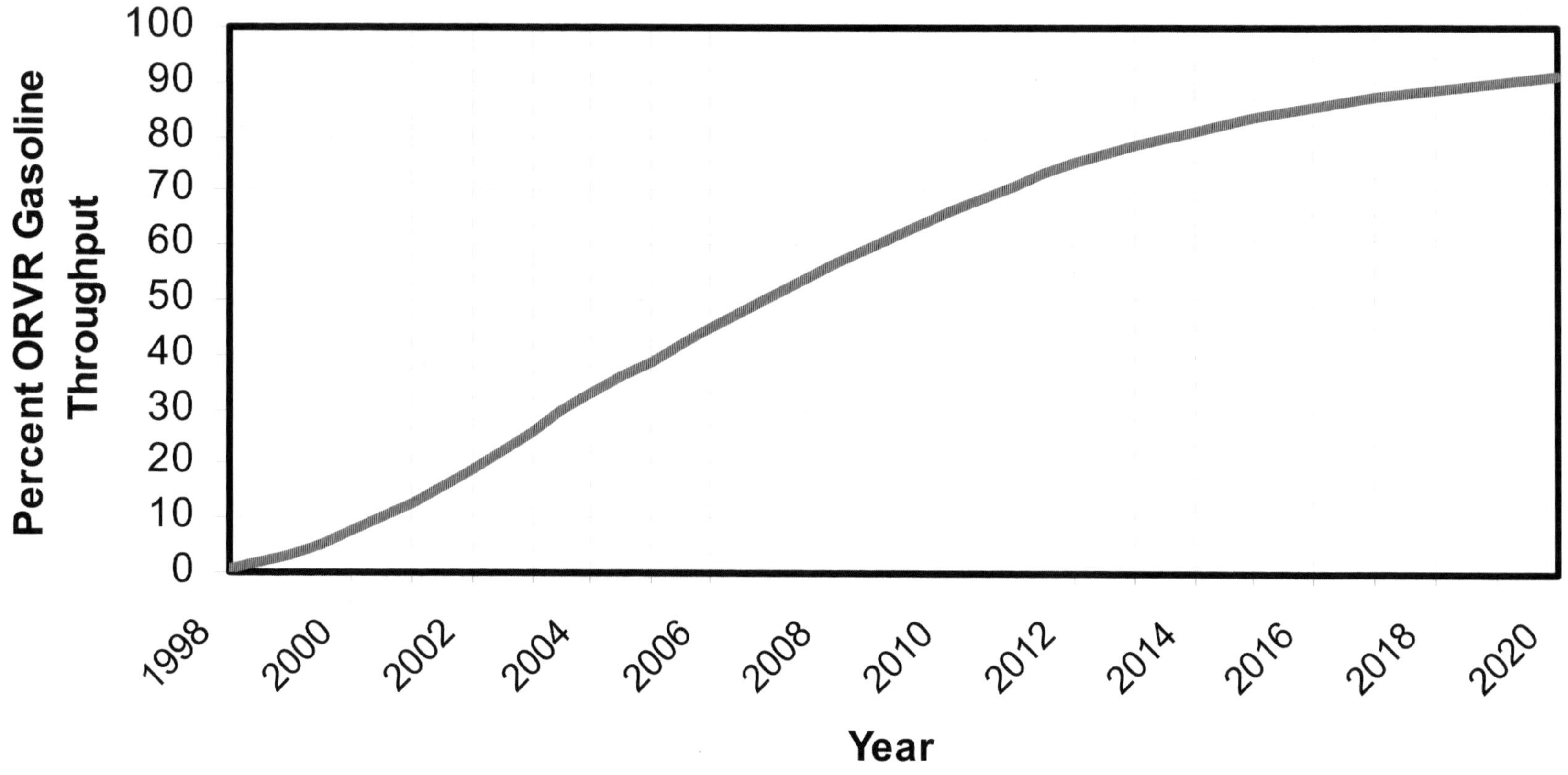

Chart courtesy of California Air Resources Board

AIChE April 2005

Gas Transport in Polymers

Feed Pressure $P_{feed,i}$ > Permeate Pressure $p_{perm,i}$

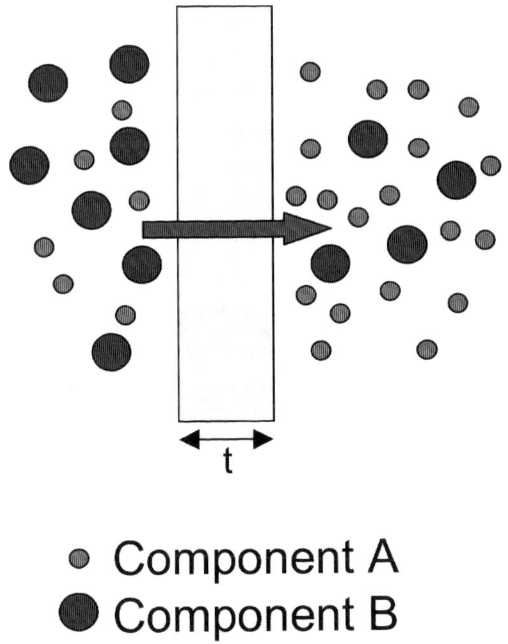

- Permeation Equation:

$$R_A = \frac{P_A\, A\, (P_{feed,A} - p_{perm,A})}{t}$$

- Permeability of A, $P_A = D_A\, S_A$

Where, D_A = Diffusion Coefficient of A
S_A = Solubility Coefficient of A

- Selectivity

$$\alpha_{A\text{-}B} = \frac{P_A}{P_B} = \left(\frac{D_A}{D_B}\right)\left(\frac{S_A}{S_B}\right)$$

Mobility Selectivity (D_A/D_B); Solubility Selectivity (S_A/S_B)

Permeability Definition

$$\text{Permeability} = \frac{\text{Volume} \cdot \text{Thickness}}{\text{Area} \cdot \text{Time} \cdot \Delta\,\text{Pressure}}$$

$$[\text{barrers}] = 10^{-10}\,\frac{\text{cm}^3\,(\text{STP}) \cdot \text{cm}}{\text{cm}^2 \cdot \text{sec} \cdot \text{cm Hg}}$$

Permeance (productivity) is higher for thinner membranes

$$\alpha_{\text{gas 1/ gas 2}} = \frac{\text{Fast Gas Permeance}}{\text{Slow Gas Permeance}}$$

AIChE April 2005

Processor Performance – Duty Cycle and Daily Throughput

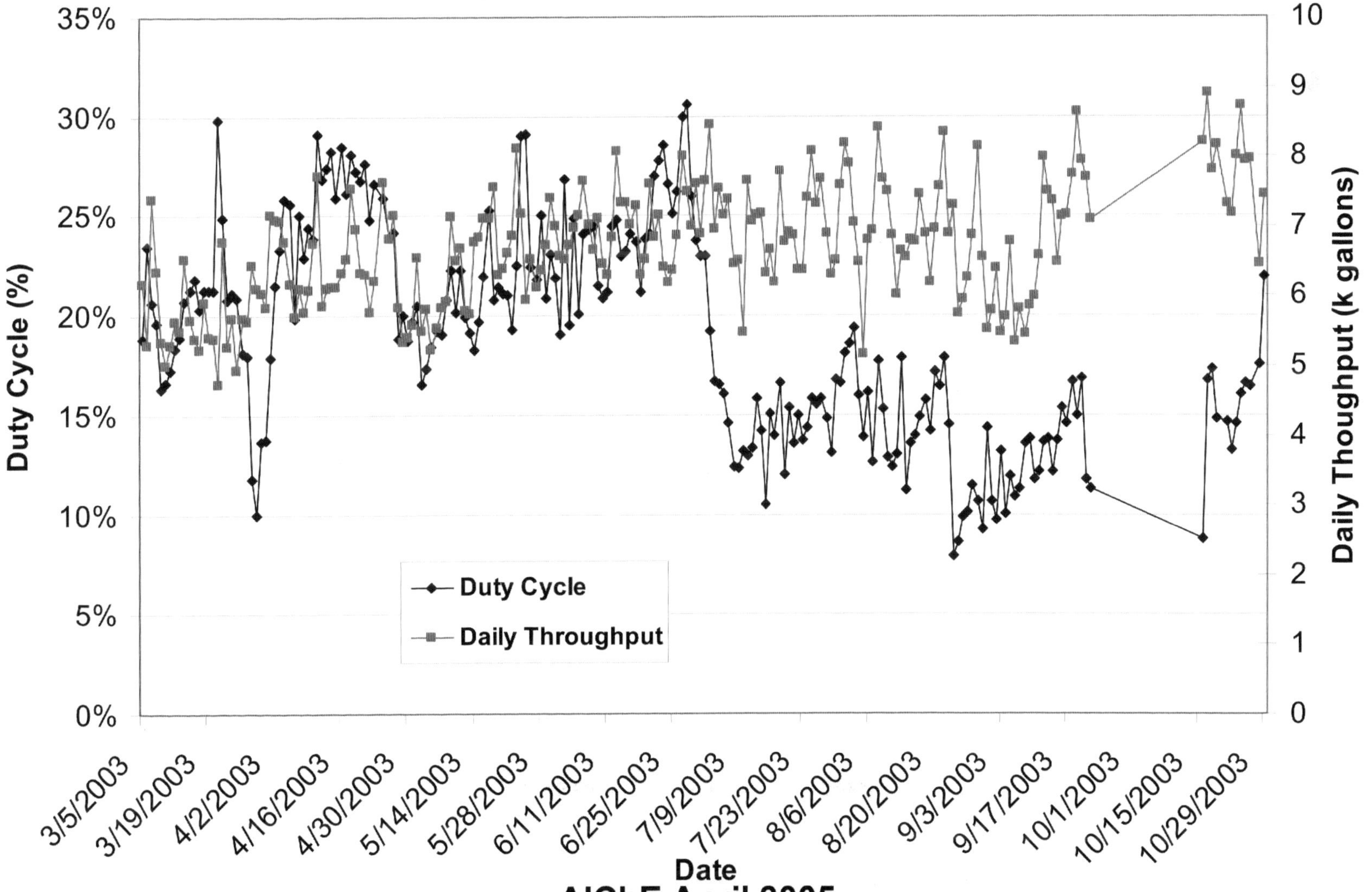

AIChE April 2005

Processor Performance - Emissions and Efficiency

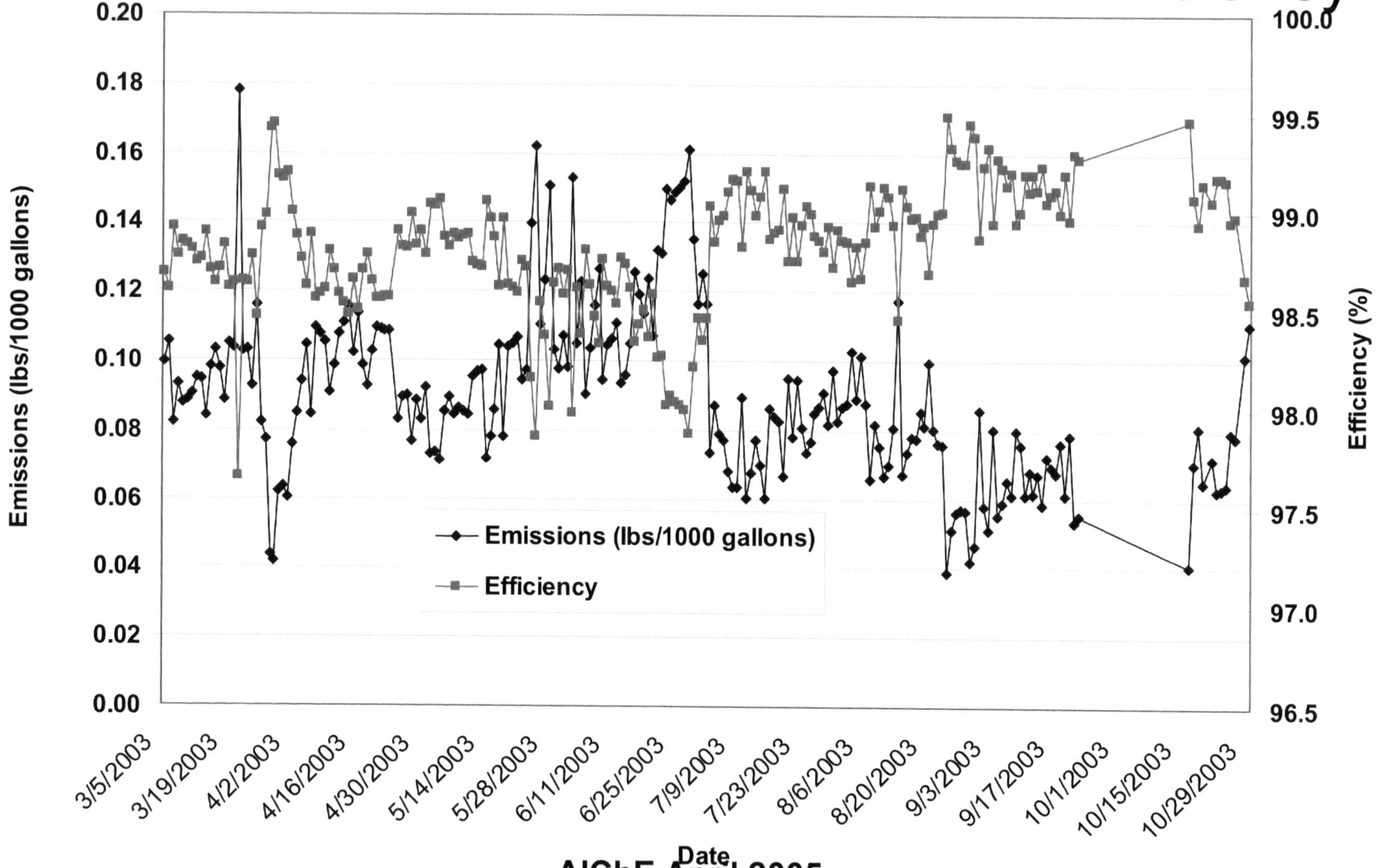

AIChE April 2005

Permeability Decreases With Molecular Size

The ratio of perms for any two gases is a good first estimate of the separation factor.

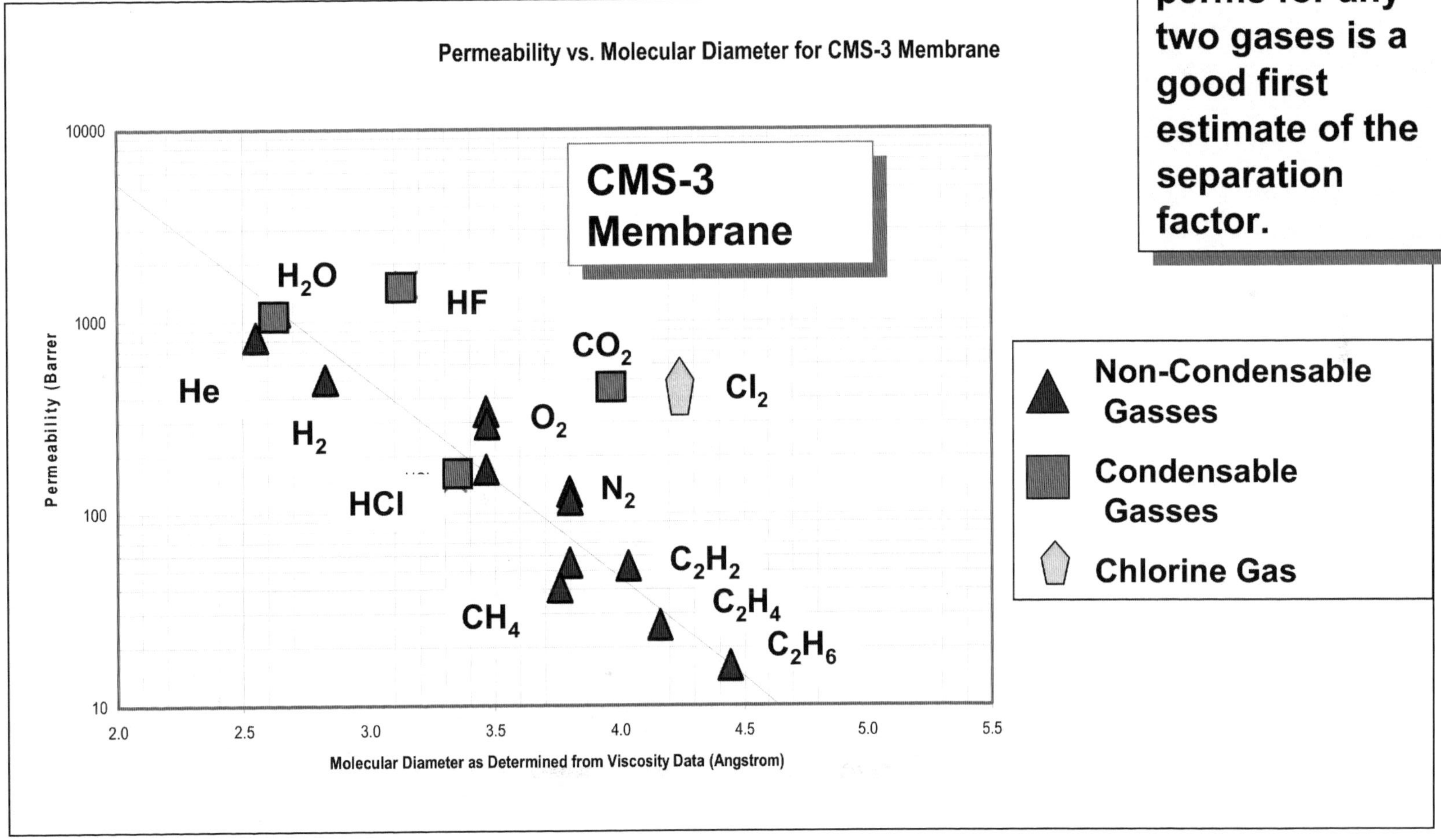

AIChE April 2005

CMS

Topics not covered in presentation

- Assist v Balance
- Details and timing of regulations, CARB's role
- Certification process
- ISD
- ORVR
- Industry structure, manufacturing and distribution chain, partners beyond VST + CMS
- Nozzle issues
- Competitive technologies
- Markets (CA, rest of USA, Worldwide)
- Other VOC applications
- Cartridge design
- Theoretical aspects of the membrane
- Results and field performance since November 2003

AIChE April 2005

Hollow Fibers as Structured Packing for Olefin/Paraffin Separation

Dali Yang, Robert S. Barbero, David J. Delvin
MST – 7, MS E549, Los Alamos National Laboratory, Los Alamos, NM 87544

Martin E. Carrera and Craig W. Colling
BP Amoco Chemical Company, 150 W. Warrenville Rd., Naperville, IL 60563-8460

E. L. Cussler
Department of Chemical Engineering and Materials Science, University of Minnesota, Minneapolis, MN 55455

Abstract:

In this study, the hollow fiber replaces the conventional tray and/or structured packing. Using a column less than 40 cm long an ~ 8% enrichment of propylene from a 30% propane/70%propylene mixture was achieved. An HTU as low as 8.8 cm was obtained. Such a low HTU has not been previously reported for propane/propylene separation. The mass transfer time was less than one second.

Introduction:

Ethylene and propylene are two of the largest commodity chemicals in the U.S. and are major building blocks for the petrochemicals industry. These olefins are currently separated by cryogenic distillation, which demands extremely low temperatures and high pressures. Over 75 billion pounds of ethylene and propylene are distilled annually in the U.S. at an estimated energy requirement of 400 trillion BTU's[1]. Many petroleum companies have been aggressively pursuing non-cryogenic alternatives to light gas separations. The largest potential area for energy reduction is the cryogenic isolation of the product hydrocarbons from the reaction by-products, methane and hydrogen. This separation requires temperatures as low as -150 °F and pressures exceeding 450 psig. This report will focus on developing a new type of packing structure - parallel hollow tubes made from plastic materials which are light weight and less costly than metal. This material will replace conventional packing materials (such as tray or structured packing) to separate olefinic and/or paraffinic mixtures from light gas byproducts with greater efficiency.

Over the past few decades many researchers have investigated the use of membranes as a contactor in distillation processes.[2,3] The main focus has been in the areas of desalinization, food industry and biological/medical applications. As material science advances, more and more polymeric materials with different geometries and morphology are becoming commercially available. Although appreciable progress has been made during the past decade, fouling of the membrane remains the primary obstacle for many of these applications. There have also been many reports on the use of reactive or selective membranes for the olefin and paraffin separation.[4-8] However, the difficulty of long term stability associated with these facilitated transport membranes has been a major obstacle. Recently, Devlin et. al. explored the possibility on separating olefin and/or paraffin gases from light gas by-products using hollow fibers, and obtained promising results.[7] Coincidently, Cussler's group reported their work on the use of non-selective membranes as structured packing to replace distillation columns for the water-isopropanol separation. They also indicated the possibility of using this technology for light hydrocarbon mixture separation.[9,10] In this report, we investigate the use of non-selective polymeric membranes as structured packing for propane/propylene separation. The height of transfer unit (HTU) for our system is as low as 8.8 cm., which is at least 5 to 10 times better than the most efficient structured packing (Sulzer 250 Y). The mass transfer coefficient is > 0.01 cm/sec. This result demonstrates the possibility of using hollow fibers or tubes as an alternate packing for olefin and paraffin separation. We have also shown that hollow fibers can help overcome flooding or loading problems offering a wider range of operating conditions. The large mass transfer area should result in increased efficiency and may ultimately reduce the operating and capitol cost compared to the conventional distillation tower.

Theory

Figure 1 (a) presents a distillation column. In a typical distillation process, the feed solution is fed at the middle section of the column, the high boiling point component is concentrated and removed from the reboiler while the low boiling point component is collected after the condenser. In order to increase the separation efficiency, a partial amount of condensed material is returned back to the column. To evaluate a distillation process it is convenient to operate under the condition of total reflux (all the condensed liquid is returned to the column) illustrated in figure 1 (b). This simplifies the analysis and allows us to determine the minimum number of stages and height of a transfer unit HTU for the separation.

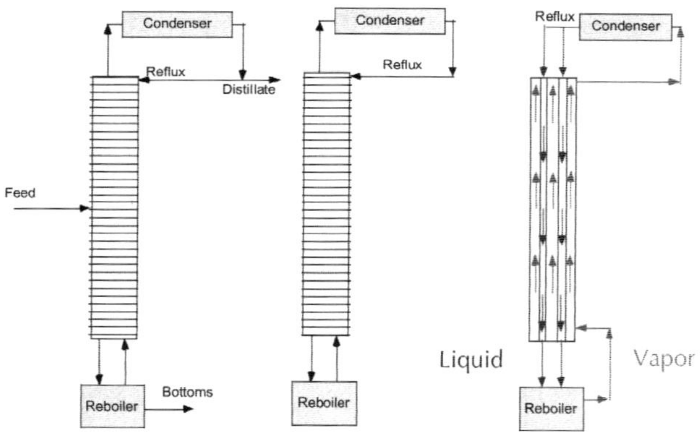

Figure 1. Flowcharts of a typical distillation process (a), a distillation process under total reflux condition (b), and a hollow fiber distillation column under total reflux condition (c).

In the hollow fiber distillation column (see Figure 1 (c)), the gas flows up the outside of fiber (shell) while the liquid counter-currently flows down inside the lumen of the hollow fiber. Using this approach we can decouple the liquid and vapor flows allowing for operation above the normal flooding limit. Under the condition of total reflux, the mass balance between the gas and liquid phase is simply:

$$G' = L', \qquad \text{Eq. 1}$$

where G′ and L′ are the molar gas and liquid flowrate in the column. In a continuous flow condition, we can assume a constant molar flow at different sections along the length of the column. For the total reflux, the operating line is simply:

$$x = y, \quad \text{Eq. 2}$$

where x and y are the mole fraction of the more volatile species in the vapor and the liquid phase, respectively. For the mass balance on a vapor phase alone, we have the following equation:

$$0 = -G\frac{dy}{dz} + K_y a(y^* - y) \quad \text{Eq. 3}$$

where K_y is the overall gas phase mass transfer coefficient, a is the mass transfer area (cm^2/cm^3), z is the distance measured from the bottom of the column, y is the mole fraction of the most volatile species in the vapor, y^* is the mole fraction in equilibrium with the liquid, and G is the gas molar flux in the shell side. This is a rate equation and the integrated form of Eq. 3 is Eq. 4:

$$l = \int_0^l dz = \frac{G}{K_y a}\int_{y_0}^{y_l}\frac{dy}{y^* - y} \quad \text{Eq. 4}$$

Eq. 4 can be written as:

$$l = HTU \bullet NTU \quad \text{Eq. 5}$$

where HTU is the height of a transfer unit defined as:

$$HTU = \frac{G}{K_y a} = \frac{v_G}{K_G a} \quad \text{Eq. 6}$$

and NTU is the number of a transfer unit given by:

$$NTU = \int_{y_0}^{y_l}\frac{dy}{y^* - y} \quad \text{Eq. 7}$$

Both HTU and NTU are important parameters. The HTU is an indication of the efficiency of the equipment while the NTU is a measure of the difficulty of the separation. A large NTU suggests a difficult separation while the small HTU is a sign of an efficient tower. Since the HTU is directly related to the mass transfer coefficient (K_G) which in turn depends on the velocity (v_G), this number changes for different operating conditions. However, typically the HTU is between 0.3 and 2.7 meters.[11,12] The NTU depends on the intrinsic properties of the materials to be separated. The propane and propylene separation is a very difficult separation due to their low relative volatility (<1.2 within a wide temperature range).

Experimental

The apparatus is shown schematically in Figure 2. There are three major components for this system: reboiler, condenser, and membrane module. There are four sampling stations for monitoring the gas composition at different sections during operation. Two heater/chiller systems are used to precisely control operation temperature. A view port mounted between the condenser and module is used to ensure the hollow fibers are filled with liquid during the experiment. A ~30% propane and propylene mixture is used in this study. All gases were 99.999% purity (Tri-gas Inc).

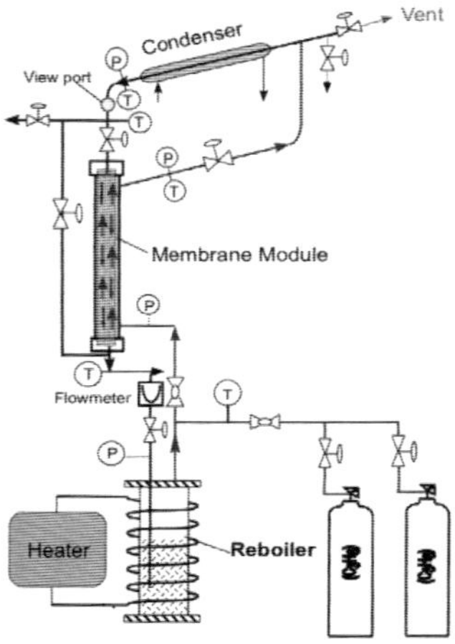

Figure 2. The flowchart of the distillation system for the olefin and paraffin separation.

HP M series micro gas chromatograph (GC) (Model number G2762A) was used to characterize the gas compositions. The RHEONIK coriolis meter (Liquid controls, LLC) with RHM 03 sensor and RHE08-transmitter was used for measuring the liquid flow. The vapor flow rate at total reflux was back calculated from the liquid flow based on the mass balance between the vapor and liquid flow. PX880 – 300 Electronic pressure transmitters (Omega) are mounted at four different sections for monitoring the pressure change in the system. T and K type thermocouples are used for monitoring the temperature. Stanford Research System model SR630 thermocouple readout with 16 channel scanner was used to acquire and digitize a 0 – 5 V voltage signal from the temperature and pressure transducers .

Several hollow fiber modules have been investigated in this study. Each module was packed with polymeric hollow fibers. Table 1 contains a summary of the module characteristics based on our own measurements.

Table 1. Summary of several modules' parameters*.

	BP - module	LANL – 3 module	LANL - 6 module	PM - module
Fiber OD (mm)	2.7	0.626	0.240	0.612
Fiber ID (mm)	1.5	1.20	0.300	0.914
Number of fiber	19	20	198	104
Mass transfer area per volume (cm^2/cm^3)	6.23	2.98	14.1	15.1

*:Active length of these modules is 36.8 cm. All of these fibers were potted inside plastic tube with 1.28 cm ID. The Celgard® hollow fibers were used in LANL - 6 module.

Results and Discussion

In general a fresh module typically required several hours to adapt to the working environment in order to reach steady state. Once it was conditioned the system needed less time to reach a steady state. Figure 3 presents the temperature and pressure profile of a typical trial conducted at near 25 °C. For this particular run, the system temperature and pressure reached a steady state within 5 hours, at which point the gas samples were withdrawn and analyzed by GC.

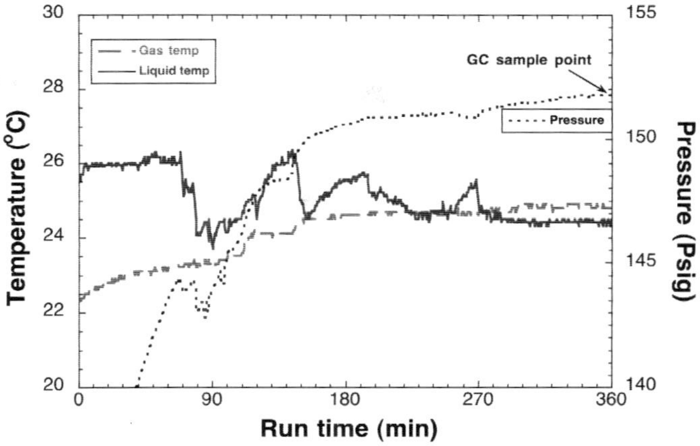

Figure 3. Temperature and pressure profiles for a typical run of BP module.

By controlling the temperature difference between the reboiler and the chiller, we are able to control the fluid velocity inside and outside of the hollow fibers. Since both liquid and vapor have their own channel to flow, the operational problems associated with loading and flooding encountered with conventional packing in a tower can be minimized. A plot of flow parameter vs. the capacity of packing materials is commonly used to indicate the operation range for different packing materials. Figure 4 presents the correlation between the flow parameter ($L/G^*(\rho_G/\rho_L)^{0.5}$– where ρ is density, the subscript of L and G indicating liquid and vapor phase, respectively) and the capacity of the modules at different operation conditions. Most of the experimental points are above the flooding line where the conventional packing materials experience flooding.

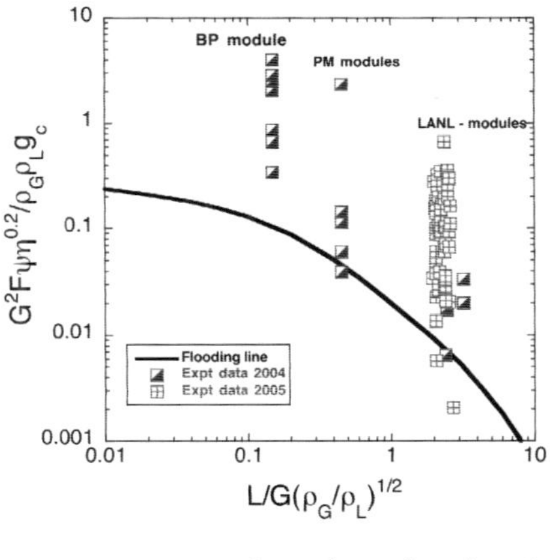

Figure 4. The correlation between flow parameter (on the abscissa) and the capacity factor of our modules (on the ordinate).

Using measured gas compositions, Eq. 7 and the V-L equilibrium data the NTU is calculated. From equation 5, we can determine the HTU for the hollow fiber modules. Figure 5 shows a plot of the experimentally obtained HTU vs. gas velocity. As is expected the HTU increases with increasing gas velocity. For the BP module, when the gas velocity is lower than 25 cm/sec, the HTU is as low as 8.8 cm, the best result reported for the propane/propylene separation.

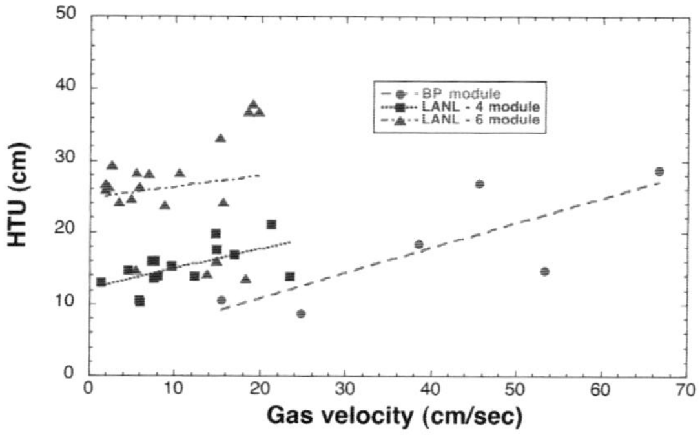

Figure 5. HTUs of BP and LANL modules as functions of vapor velocity.

Conclusion

In this report, we demonstrated the possibility of using hollow fiber as structured packing for the olefin/paraffin separation. Due to its large surface area, a low HTU (<10 cm) as compared to conventional packing is achieved. Furthermore, the hollow fibers, allow the liquid and vapor to flow counter-currently and separately on the inside and outside of hollow fiber. The flooding and loading problems common for the conventional packing materials are minimized. Using low cost and lightweight polymeric materials to replace metal as structured packing can decrease capital cost.

Acknowledgement: DOE Energy Efficiency and Renewable Energy, OIT, Industrial Technology Program funds this project, which the authors are thankful. John H. Burban of Porous Media Corporation (St. Paul, MN) provided the PM – module.

Reference:

(1) Eldridge, R. B. *Ind. Eng. Chem. Res.* **1993**, *32*, 2208-2212.
(2) Lawson, K. W.; Lloyd, D. R. *J. of Membrane Science* **1997**, *124*, 1-25.
(3) Cabassud, C.; Wirth, D. *Desalination* **2003**, *157*, 307-314.
(4) Safarik, D. J.; Eldridge, R. B. *Ind. Eng. Chem. Res.* **1998**, *37*, 2571-2581.
(5) Padin, J.; Yang, R. T. *Chemical Engineering science* **2000**, *55*, 2607-2616.
(6) Sungpet, A.; Way, J. D.; M.Thoen, P.; Dorgan, J. R. *J. of Membrane Science* **1997**, *136*, 111-120.
(7) N. Calamur, M.E. Carrera, D. J. Devlin, T. Archuleta, US Patent *6,039,792*, 2000
(8) Hsiue, G. H.; Yang, J. S. *J of Membrane Science* **1993**, *82*, 117-128.
(9) Zhang, G.; Cussler, E. L. *J of Membrane Science* **2003**, *215*, 185-193.
(10) Zhang, G.; Cussler, E. L. *AIChE Journal* **2003**, *49*, 2344.
(11) Prasad, R.; Sirkar, K. K. *AIChE Journal* **1988**, *34*, 177.
(12) Cussler, E. L. *Diffusion*; 2nd ed.; Cambridge University Press: New York, 2002.

22f Distillation Column Flooding Predictor; A Cost Effective Distillation Column Debottlenecking Technology for Petrochemical Industry

George E Dzyacky

The demand for gasoline in the US has grown more than 20 percent in the last twenty years while refining capacity has dropped 10 percent. Trying to solve this problem by adding capacity at existing refineries is becoming harder to do because of more stringent environmental regulations and the existence of a complex and open-ended permitting process. One solution is to make the existing processes more energy efficient and productive.

Working under a Cooperative Agreement from the U. S. Department of Energy, Chemicals Industry of the Future, 2ndpoint, LLC has developed a low-cost, control technology that enables refineries to increase distillation column throughput. Recent tests conducted at The University of Texas at Austin indicate the pattern recognition technology increases distillation column through put by as much as six percent.

The company's patented Distillation Column Flooding Predictor offers a low-cost, low-risk, distillation column de-bottleneck alternative to capital expansion projects. The technology identifies the onset of flooding in distillation columns, allowing refiners to operate columns nearer their true hydraulic limit, thereby increasing throughput and avoiding flood.

The benefit of the Distillation Column Flooding Predictor arises out of the technology's ability to control distillation columns at their hydraulic limit – a constraint that current distillation control technologies cannot recognize let alone control to.

23a A New Corrosion Management Technology for the Chemicals Industry

A D Pelton, A L Young, I G Wright, Randy C John, W T Thompson

A novel technology is being developed to assess, predict and manage corrosion of a wide variety of metals and alloys used to fabricate equipment found in many different types of process conditions. The presentation will be in two parts: corrosion related to corrosion by oxidation, sulfidation, sulfidation/oxidation, carburization and corrosion by HCl/Cl2 gases, metals dusting and cyclic oxidation. Application examples of the technology to solving real engineering problems and likely energy savings will be discussed. This technology currently archives more than 13 million hours of exposure data for 90 commercial alloys exposed to conditions found in many different types of petrochemical, petroleum refining, pulp/paper processing and power generation process conditions.

23b Flow Mapping of Gas-Solid Riser

M H Al-Dahhan, M P Dudukovic, S Bhusarapu

Gas-solid risers are extensively used in many industrial processes for gas-solid reactions (coal combustion and gasification, etc.) and for catalytic gas phase reactions (fluid catalytic cracking, butane oxidation to maleic anhydride, etc.). Risers pose considerable challenges because it is not yet possible to produce ab initio predictions of their complex fluid dynamics, which impairs reactor modeling. In the Chemical Reaction Engineering Laboratory (CREL) at Washington University, Computer Automated Radioactive Particle Tracking (CARPT) has been developed and extensively utilized for measuring in a non-invasive manner the velocities, turbulent parameters and 3D flow mapping in multiphase flows. In this study, solids flow dynamics has been investigated in "cold-flow" circulating fluidized bed risers using non-invasive flow techniques – CARPT and Computed Tomography (CT). Comparative and symbiotic analyses of the results obtained from CARPT and CT have been used to develop a coherent picture of the solids flow field.

American Institute of Chemical Engineers
3 Park Avenue
New York, NY 10016-5991